84/-

ARMAGH COUNTY LIBRARY

TIME ALLOWED FOR READING

This book is issued for **14** days. If kept beyond this period a fine of twopence per week, or part of a week, will be incurred.

29 MAY 1971	3 APR 1977 05	
12 JUN 1971 05		
22 JUL 1971 05		
4 OCT 1972	BROWNLOW LIBRARY	
	TULLYGALLY	
	CRAIGAVON	
7 JUN 1974		
-2. AUG. 1974	15. SEP. 1977	
20. SEP. 1974		
5 APR 1975		
-1. JUL. 1975		
1 FEB 1976		
LURGAN		

THE LUNAR ROCKS

The near side of the moon. The dark areas are the maria, the brighter areas the highlands; the prominent crater with the extensive ray system in the southern hemisphere is Tycho. (Mount Wilson Observatory photograph.)

THE LUNAR ROCKS

BRIAN MASON
WILLIAM G. MELSON

Smithsonian Institution
Washington, D. C.

WILEY-INTERSCIENCE
A Division of John Wiley & Sons, Inc.
New York • London • Sydney • Toronto

184895

523.34

Copyright © 1970, by John Wiley & Sons, Inc.

All rights reserved. No part of this book may be reproduced by any means, nor transmitted, nor translated into a machine language without the written permission of the publisher.

Library of Congress Catalogue Card Number: 73-129659

ISBN 0 471 57530 5

Printed in the United States of America

10 9 8 7 6 5 4 3 2 1

PREFACE

The lunar rocks represent a unique scientific adventure and an intellectual challenge of the first magnitude. From the data to be extracted from them, we expect to reconstruct the story of the origin and evolution of the Moon, which will also provide significant information about the early history of the Earth and the entire solar system. In addition, the Moon's surface contains a record of the radiation received from the Sun and from galactic sources over hundreds of millions of years, preserved because of the absence of weathering. For those scientists privileged to be selected to investigate the Apollo 11 collections, the sight on television of astronaut Neil Armstrong bagging the first lunar sample had a unique significance—the realization that the man-years of thought, training, and experimentation in preparation for the lunar samples were now to be put to the ultimate challenge.

The results of this challenge were disclosed at a meeting in Houston, Texas, on January 5–8, 1970, when 140 principal investigators and many of their collaborators presented and discussed the data they had obtained. These results were reported in an abbreviated form in the January 30 issue of *Science*, and at greater length in a special supplement of *Geochimica et Cosmochimica Acta*. Being directly involved in this enterprise, we were naturally anxious to learn to what extent our results were corroborated by other investigators, to what extent there was conflict of evidence, what additional information was forthcoming, and most important, how all this information could be collated and interpreted to provide a coherent and internally consistent account of lunar history.

The lunar rocks are certainly the most intensively and extensively studied materials in the history of science. Already the plethora of data is dismaying for anyone wishing to obtain a comprehensive review of the results obtained. Our own interest in obtaining such a review prompted us to write this book—an attempt to provide a concise and coherent account of the scientific effort on the lunar samples and the interpretation of the results.

The book begins with a brief review of what has been learned about the moon by telescopic examination and by unmanned spacecraft. Chapter 2 presents a summary of the organization of the Apollo missions and the plans for future landings. The major part of the book examines lunar mineralogy, lunar petrology, and lunar geochemistry, our own fields of specialization and the areas in which the factual data are most extensive. Lunar mineralogy, while limited so far to about 30 species, shows many similarities to that of terrestrial basalts, but also some significant differences, plus the presence of at least two new minerals. Lunar petrology is essentially an igneous-rock petrology, with the addition of complex breccias evidently produced by the impact events recorded in the cratered surface of the Moon. Most of the rocks so far collected are basaltic in mineralogical composition and texture, but nonbasaltic types have also been found in small amount. The lunar geochemistry comprises the measurement of all the chemical elements in the lunar samples, and this has shown an intriguing pattern of elemental depletions and enrichments relative to terrestrial rocks, meteorites, and probable cosmic abundances. The final chapter presents the application of all this information to the elucidation of the grand problems of the origin and history of the Moon.

We have endeavored to write this book in such a way that it will appeal not only to the professional scientist, but also to the interested student and layman. With this in mind we have avoided technical jargon as far as possible and have tried to provide adequate explanations of unfamiliar terms. We hope that the book gives some indication of the enormous stimulation that the lunar exploration program has provided to the scientific community, in this country and abroad.

Our own researches have been supported in part by grants and contracts from the National Aeronautics and Space Administration, and this support is gratefully acknowledged. We are also greatly indebted to numerous colleagues, whose formal communications and informal conversations and discussions have contributed so much to this book. Special thanks are due to Richard A. Allenby, John A. O'Keefe, and Norman Watkins, who read and commented on various parts of the manuscript.

BRIAN MASON
WILLIAM G. MELSON

Washington, D.C.
March 1970

CONTENTS

1. **Introduction: Pre-Apollo** — 1
2. **Apollo** — 14
3. **Lunar Mineralogy** — 32
4. **Lunar Petrology: The Igneous Rocks** — 51
5. **Lunar Petrology: The Fines and Microbreccias** — 80
6. **Lunar Petrology: Comparisons with Terrestrial Rocks, Meteorites, and Tektites** — 100
7. **Lunar Geochemistry** — 116
8. **Implications for Lunar History** — 155

   References — 163

   Index — 171

THE LUNAR ROCKS

CHAPTER 1

INTRODUCTION: PRE-APOLLO

By any standards the decade of the 1960s was surely unique for science as a whole and for geology in particular. It began with a first view of the far side of the Moon and it ended with the first look at the material from which the Moon is made. In the intervening ten years the Moon was intensively mapped by remote-controlled spacecraft, both Russian and American.

It is worth recalling that at the beginning of the decade the knowledge of our nearest neighbor in space rested essentially on the exploitation of Galileo's pioneering use of the telescope in 1610 to examine the surface of the of the Moon. Galileo provided the first topographic description of the Moon. He observed the great dark smooth areas that he called *maria* or seas, bordered by high ranges of *mountains* (Fig. 1-1). He saw areas pitted with ring-shaped formations, which he called *craters*. These areas, which appear white in contrast to the dark maria and rise above them, are now known as the *terrae* or *highlands* (Fig. 1-2). Later, additional topographic features were described and named. Of these, *rilles* and *rays* are particularly intriguing. Rilles (Fig. 1-3) are long (up to hundreds of kilometers) narrow channels that occur in large numbers, particularly on the maria. Their canyonlike appearance has stimulated much speculation that they were formed by running water; however, this explanation now seems improbable. Rays (frontispiece) are narrow bright streaks extending radially for hundreds of kilometers from some of the craters, probably material ejected from within the crater at the time of its formation. *Wrinkle ridges* are sinuous elevations occurring on the maria, irregular in form and quite variable in width and length and rarely exceeding a few hundred feet in height. They have been interpreted as volcanic features resulting from the extrusion of lava along fractures. *Crater chains,* as the name implies, are linear groups of small craters, the craters being strung along the line like

Fig. 1-1. Mare Imbrium, bordered on the southeast by the Apennine Mountains. The prominent rayed crater at the southern margin of the photograph is Copernicus (90 km in diameter); the smaller crater with the prominent central peak on the south rim of Mare Imbrium is Eratosthenes. (Mount Wilson Observatory photograph.)

Fig. 1-2. The southwest quadrant of the Moon, showing the cratered highlands on the eastern and southern side of Mare Nubium; the Straight Wall is right of center. (Mount Wilson Observatory photograph.)

4   THE LUNAR ROCKS

Fig. 1-3. Rima Hadley (Hadley's Rille), the sinuous depression illustrated above, is one of the most conspicuous lunar rilles when viewed with Earth-based telescopes. Its southern end (bottom) is a very deep elongate depression, interpreted by some as its source. The rille meanders through mare material in a northeasterly direction, approximately parallel to the margin of the Apennine Mountains, which rise thousands of feet above the adjacent mare surface. The prominent circular crater in the center of the photograph is Hadley C, 5.5 km across. (NASA Orbiter V photograph.)

beads on a string; this linear distribution suggests a common origin, and they are inferred to represent volcanic eruptions along lines of weakness. *Faults* are dislocations of the lunar crust, manifested by linear scarps and dislocations; the most prominent is known as the Straight Wall, a scarp about 110 km long and up to 300 m high in the southeastern part of Mare Nubium (Fig. 1-2).

The abundance of craters is certainly a most remarkable feature of the Moon's surface; their origin, whether by volcanism or by the impact of large meteorites, has been argued for many years. These craters range in size from Bailly (294 km in diameter) and Clavius (235 km in diameter) down to small pits. Craters are usually surrounded by a raised rim, and a significant generalization is that the volume of material in the rim is approximately equal to the volume of the interior depression. Many craters have have central peaks or mountains (e.g., Copernicus, Fig. 2-10); others have smooth flat floors, possibly indicating that they have been partially flooded with lava. Cratering was evidently a long-continued process, as is demonstrated by Fig. 1-4, which shows an old flooded crater with many younger craters superimposed upon it. The youngest craters are the brightest and show sharp blocky rims, prominent rays, and fields of satellite craters around them (Fig. 1-5).

From Galileo onward improvements in instruments and techniques led to improved resolution (ultimately to distinguishing details down to a few hundred meters across), and thus to better pictures and maps of the front side of the Moon. (Since the Moon rotates on its axis once in its period of revolution about the Earth, only one side—more precisely 59% of its surface because of orbital variations—is visible to an Earth-bound observer.) The best telescopic maps of the Moon have been the 1:1,000,000 sheets prepared by the Aeronautical Chart and Information Center of the United States Air Force. One significant aspect of these maps is that they represent perhaps the first scientific attempt to establish contours (300-m interval) on the lunar surface with any degree of accuracy.

Using these sheets as the topographic base, the U.S. Geological Survey adapted standard photogeologic techniques to prepare the corresponding 1:1,000,000 geologic maps. Using the geologic law of superposition, the chronologic relationships of many of the lunar surface features can be unambiguously determined from telescopic observations. On this basis, the following broad stratigraphy of the lunar surface was established (Shoemaker and Hackman, 1962):

Period	*Events*
Corpenican	Formation of ray craters
Eratosthenian	Formation of large craters whose rays are no longer visible
Imbrian	Period of mare formation
Pre-Imbrian	Formation of the lunar highlands

From telescopic examination of the Moon, the following conclusions seem valid:

1. The pre-mare history resulted in an extremely rough, heavily cratered surface.
2. The mare material covered much of the rough pre-mare surface and is very much smoother on a gross scale.

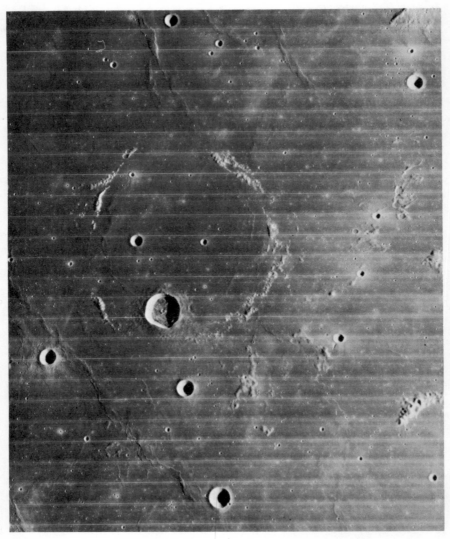

Fig. 1-4. A mare area in Oceanus Procellarum, showing an extreme range in crater size and freshness; the broken ring (100 km in diameter) of bright ridges is probably the rim of an old (pre-mare) crater that has been almost completely buried by mare material. It has also been interpreted as a ring dike by O'Keefe et al., 1967. The mare surface is pitted with post-mare craters; the large one near the southern edge of the ring is Flamsteed, about 20 km in diameter. (NASA Orbiter IV photograph.)

INTRODUCTION: PRE-APOLLO 7

Fig. 1-5. An oblique view of Kepler, a Copernican-age crater, 32 km in diameter. The hummocky ejecta blanket, radial ridges, and field of satellite craters are typical of relatively young craters. Kepler and its related features are superposed partly on mare and partly on rugged highlands that may be part of the outer rim of Mare Imbrium. (NASA Orbiter III photograph.)

3. The post-mare processes were much reduced in intensity in comparison to the earlier periods of lunar history, leaving the maria little changed since their formation, at least on a large scale.

Many inferences have been made about the nature of the lunar surface materials by studies of reflected light. Wide variations were noted in the brightness of lunar surface materials, but even the brightest could be accounted for by relatively dark materials. Many terrestrial materials, such as light-colored volcanic ash, granitic rocks, and most kinds of sandstones could be ruled out. All that could be said with some certainty was that (1) most of the lunar surface was covered with porous, relatively fine-grained, presumably fragmental materials, and (2) these materials had very low albedos. No unique mineralogy or composition could be determined, because a wide variety of substances may have identical albedos. Still, many investigators referred to a volcanic ashlike veneer and to the possible existence of a sizable basaltic component.

Some observers viewed certain darker and lighter areas of the maria as representing great individual lava flows. They further inferred, from the differences in crater densities, that eruptions in some regions were widely separated in time. These lunar flows were envisioned as being similar to the great terrestrial lava outpourings of flood basalt fields, such as those of the Columbia River Plateau. Significantly absent were large constructional features corresponding to terrestrial andesitic stratovolcanoes, which owe their form to central outpourings and explosions of very viscous lava. Suspected lunar lava flows appeared to have margins with very slight elevations above the surrounding surface. This was attributed to very low lava viscosities (Baldwin, 1963).

Prior to Apollo, of course, there were no direct age determinations of lunar materials. Thus only the above-outlined relative ages for various parts of the lunar surface had been established. Nevertheless, most scientists believed that most major lunar surface features were extremely ancient. Indeed, some postulated that most lunar rocks would be older than the oldest rocks on Earth, which date back at least 3 billion years. The Moon was often thought of as a source of many clues about the early history of the solar system, and that perhaps rocks dating back to the supposed time of formation of the Moon and Earth, 4.5 billion years ago, would be found. Further, it was expected that these very ancient rocks would be found in the highlands, as the oldest regions of the Moon.

Certain characteristics of the Moon, such as its distance from the Earth, have been measured and known for many years. Some of these characteristics are:

Mean distance from the Earth	384,402 km
Diameter	3476 km (0.2723 that of Earth)
Volume	2.199×10^{10} km^3 (0.0203 that of Earth)
Mass	7.353×10^{25} g (0.0123 that of Earth)
Density	3.34 g/cm^3 (Earth is 5.517 g/cm^3)
Surface gravity	0.165 that of Earth

A new era in lunar exploration began with the launching of the Russian Luna 3 spacecraft on October 4, 1959. This spacecraft orbited the Moon and transmitted back to Earth a series of photographs from which the Russians were able to prepare a map of the far side of the Moon, never before seen. The far side shows the same variety of topographic features as the near side, except that there is a dearth of large maria.

During the first part of the 1960s NASA developed the Ranger program for close photography of the lunar surface. The early Rangers failed for a variety of reasons, but Rangers 7, 8, and 9 (Fig. 1-6), which flew in 1964–1965, were brilliantly successful, sending back to Earth 17,225 closeup photographs of the Moon. They began telecasting pictures when they were about 20 min away from the Moon and continued to transmit until crashing into the surface. As a result, features as small as a foot across were made visible to man for the first time—a thousand times greater resolution

Fig. 1-6. Map of the near side of the Moon, showing some of the major features and the locations of Ranger impacts (R7, R8, R9) and Surveyor landings (S1, S3, S5, S6, S7). (From W. M. Kaula, *An Introduction to Planetary Physics*, John Wiley & Sons, New York, 1968.)

than could be achieved by the best telescope operating under optimum conditions.

On February 2, 1966, the Russian Luna 9 made a soft landing on the lunar surface and transmitted photographs back to Earth. To the great relief of engineers responsible for designing manned landing craft, the lunar surface proved to be quite stable and capable of supporting a spacecraft. On April 3, 1966, Luna 10 orbited the Moon, carrying with it a gamma-ray spectrometer. The analysis of the gamma-ray spectrum from the lunar surface (Vinogradov et al., 1966) provided the first significant geochemical data on the Moon. The abundances of potassium, thorium, and uranium indicated by the spectrum were comparable with those in terrestrial basalts and inconsistent with a surface of granitic composition, of ultrabasic composition, or of the composition of chondritic meteorites.

At about the same time the NASA Surveyor program was in operation. This program was designed to accomplish a series of soft landings with automated spacecraft capable of acquiring and transmitting scientific and engineering measurements from the lunar surface. Surveyor I landed on June 2, 1966, and was followed at intervals by other craft through Surveyor VII, which landed on January 10, 1968. Surveyor II crashed on the Moon, Surveyor IV landed successfully but transmitted no data; all the other Surveyors were eminently successful. Surveyor I landed on a flat surface inside a 100-km crater in Oceanus Procellarum; III, in the interior of a 200-m crater, also in Procellarum (this spacecraft was later visited by the crew of Apollo 12); V, in a 10-m crater in Mare Tranquillitatis; VI, on the flat surface of Sinus Medii; VII, on the exterior rim of the large crater Tycho (Fig. 1-6). All except the last landed in mare areas in the Moon's equatorial belt, primarily to explore potential Apollo landing sites. These tasks having been successfully accomplished, Surveyor VII was sent to an area of primary scientific interest, the rugged highland terrain around the prominent ray crater Tycho.

The Surveyor spacecraft provided an enormous amount of scientific and engineering information on their landing sites. The most exciting data, however, for the geologist and geochemist were the first chemical analyses of lunar materials. These analyses were obtained by a "black box" attached to Surveyors V, VI, and VII. This "black box" was a metal cube, about 6 in. on the side, which on command was gently swung out and lowered to the lunar surface. Within the box was a source of alpha particles (curium 242), and detectors for alpha particles back-scattered from the lunar surface and for protons generated in the lunar surface materials by the alpha particles. This remarkable instrument was designed by A. L. Turkevich of the University of Chicago and his co-workers. The resultant spectra contain quantitative information on all the major elements in the materials analyzed, except hydrogen, helium, and lithium. The results of these analyses are given in Table 1-1.

Of especial interest is the Surveyor V analysis, since this was in Mare Tranquillitatis, not far from where the Apollo 11 collections were made. The laboratory analysis of a sample of Apollo 11 fines is included in Table 1-1 to show the consistency between the telemetered analysis and the laboratory figures. This comparison gives confidence in the alpha-scattering analyses and indicates that compositions at different places in the same mare may be quite similar, whereas different maria may have distinctive compositions.

Turkevich et al. (1969) compared their Surveyor V analysis with the known compositions of oceanic basalts and the eucrites, a class of achondritic meteorites (Fig. 1-7). The match is clearly quite good for the major elements oxygen, magnesium, aluminum, silicon, calcium, and iron. However, the Surveyor V analysis is notably higher in titanium than either the basalts or the eucrites, and its sodium content is much less than the basalts but similar to that of the eucrites.

The alpha-scattering analyses confirmed the Russian conclusion from their gamma-ray experiment that at least parts of the lunar surface are probably of basaltic composition. They showed that different maria are similar but not identical in chemical composition, and that the highlands have a related composition, but one which is clearly lower in iron and higher in aluminum, that is, probably more feldspathic.

The NASA Orbiter program was a series of spacecraft designed to orbit the Moon and return pictures and other telemetered information. The first was launched on August 10, 1966, and four more followed. Complete photographic cover of both the near and far side of the Moon was ob-

Table 1-1. Chemical Analyses (wt %) at the Surveyor Landing Sites, Derived from the Alpha-scattering Experiments, and an Analysis of Apollo 11 Fines (Peck and Smith, 1970); V and Apollo 11, Mare Tranquillitatis; VI, Sinus Medii; VII, Rim of Crater Tycho

	V	Apollo 11	VI	VII
SiO_2	46.4	42.3	49.1	46.1
TiO_2	7.6	7.3	3.5	0.0
Al_2O_3	14.4	14.1	14.7	22.3
FeO	12.1	15.8	12.4	5.5
MgO	4.4	7.9	6.6	7.0
CaO	14.5	12.0	12.9	18.3
Na_2O	0.6	0.5	0.8	0.7

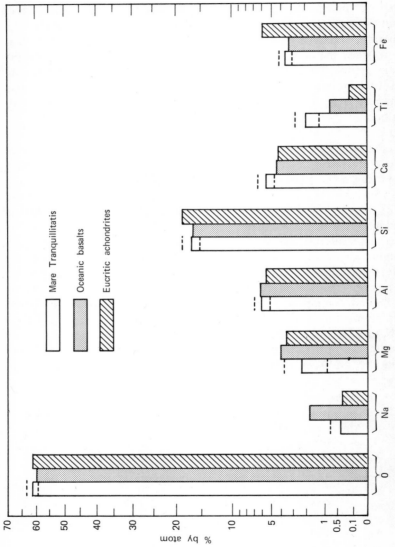

Fig. 1-7. Comparison of the alpha-scattering analysis of lunar surface material in **Mare Tranquillitatis** with analyses of oceanic basalts and eucritic meteorites. The values are plotted on a square-root scale with estimated errors indicated for the lunar results. (Turkevich et al., 1969.)

tained, and a large amount of geophysical data. Analysis of the gravity data led to the surprise discovery of mascons: localized mass concentrations beneath the Moon's surface. About 13 mascons have been identified (Kaula, 1969), most on the near side but at least one on the far side, and they are localized beneath the maria; however, not all maria have associated mascons. The largest mascons are under the circular maria: Imbrium, Serenitatis, Crisium, Nectaris, Aestuum, Humorum, Humboltianum, Orientale, and Smythii. The size of a mascon is roughly proportional to the area of the mare where it is located.

CHAPTER 2

APOLLO

Program Apollo was planned to carry out manned landings on the surface of the Moon. As finally developed (Fig. 2-1) the spacecraft in lunar trajectory consists of three units: the command module, the service module, and the lunar excursion module (LM). The command module serves basically as crew quarters for three men for launch, lunar trajectory, lunar orbit, return trajectory, reentry, and landing. The service module contains the propulsion units required for midcourse maneuvers and reentry braking for lunar orbit, and the primary life support and power equipment; it is jettisoned before final reentry into the Earth's atmosphere. The LM (Fig. 2-2) consists basically of two sections, the descent stage and the ascent stage. The descent stage comprises the landing gear and serves as a launch base for the ascent stage when it blasts off from the moon surface. The pressure cabin for crew quarters is in the ascent stage. The LM crew consists of two men; the third member of the Apollo crew remains in the command module in lunar orbit. After the ascent stage rejoins the command module the LM crew transfers to the command module with instruments and collections, and the ascent stage is then abandoned.

The early Apollo missions were unmanned excursions in Earth orbit to test the various components. The first manned flight was Apollo 7, which flew in Earth orbit October 11–22, 1968. Apollo 8 followed soon after, on December 21–27, 1968. One of the prime tasks of this mission was selected scientific photography of the lunar surface to provide information on the approach topography and landmarks for the early Apollo landings, and the broad structure and characteristics of the lunar surface. In addition, extensive photography of the lunar far side was possible; although almost all the far side had been photographed by the Orbiter flights, this photography generally was made with the spacecraft relatively far from the Moon, thereby limiting the possible resolution. Thus Apollo

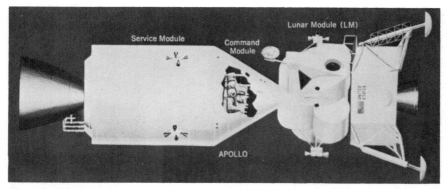

Fig. 2-1. The Apollo spacecraft: service module, command module, and lunar module.

pictures of the far side would have much better resolution than existing pictures.

These tasks were carried out with a high degree of success. The spacecraft was placed in an elliptical lunar orbit at 69 hr 8 min after liftoff. After flying two elliptical orbits of 168.5 by 60 mi with an inclination of 12° to the lunar equator, the spacecraft was placed in a nearly circular orbit of 59.7 by 60.7 mi, in which it remained for eight orbits. At 89 hr 19 min, transearth injection was performed from behind the Moon. A nearly flawless mission was completed on December 27 by splashdown in the Pacific Ocean.

Apollo 9 (March 3–13, 1969) was a test flight of the complete Apollo spacecraft, including the lunar module, in Earth orbit; this was the first manned flight of the LM. This was soon followed by Apollo 10 (May 18–26, 1969), which was a manned lunar mission to evaluate LM performance in the lunar environment. The LM was manned and flown to within 50,000 ft of the Moon's surface.

All of these exploratory missions culminated in Apollo 11, the first manned lunar landing. For a geologist the excitement of watching on television as astronaut Neil Armstrong stepped onto the lunar surface and scooped up that first sample of lunar rocks was a unique experience. Apollo 11 was launched on July 16, 1969, and the LM landed in the southwestern part of Mare Tranquillitatis on July 20, approximately 25 km southeast of the landing site of Surveyor V. The astronauts Neil Armstrong and Edwin Aldrin were on the lunar surface for 21 hr, of which almost 3 hr were spent in extravehicular activity (EVA). The EVA was carefully programmed to accomplish the following scientific objectives, in order of priority:

16 THE LUNAR ROCKS

Fig. 2-2. The lunar module on the surface of the Moon (Apollo 11). Astronaut Edwin Aldrin is removing scientific equipment from a storage bay in the descent stage, which supports the ascent stage containing the crew cabin. (NASA photo 69-H-1395.)

1. To collect early in the EVA a sample, called the contingency sample, of approximately 1 kg of surface material to insure that some lunar material would be returned to Earth.
2. To fill rapidly one of the two sample-return containers with approximately 10 kg of material, called the bulk sample, to insure the return of an adequate amount of material to meet the needs of the principal investigators.
3. To deploy three experiments on the lunar surface: (a) a passive seismometer to study lunar seimic events, the Passive Seismic Experiment

Package (PSEP); (b) an optical corner reflector to study lunar librations, the Laser Ranging Retroreflector (LRRR); (c) a solar-wind composition (SWC) experiment to measure the types and energies of the solar wind on the lunar surface.

4. To fill the second sample-return container with carefully selected material within the local geologic context, to drive two core tubes into the surface, and to return the tubes with the stratigraphically organized material, called the documented sample.

5. To obtain closeup high-resolution stereographic photographs of the undisturbed lunar surface.

These objectives were all satisfactorily accomplished, although the time available for collecting the documented sample was extremely short, limiting the amount of documentation such as photographing individual specimens in place.

The astronauts traveled in all about 750–1000 m around the landing site, the farthest single traverse being to a 33-m-diameter crater 60 m east of the LM. At the landing site, the surface was made up of unsorted fragmental debris ranging in size from the finest dust to rocks up to 0.8 m across (Fig. 2-3). This debris layer, known as the lunar regolith, was porous and weakly coherent at the surface and graded downward into similar but more compacted material. The bulk of the regolith consists of fine particles (frequently referred to as the lunar soil, a convenient term but perhaps misapplied, since soil on Earth is the product of a complex of organic and inorganic reactions unknown on the Moon).

The contingency sample, weighing about 1 kg, was scooped up from the surface close to the LM and placed in a teflon bag at the beginning

Fig. 2-3. The Apollo 11 landing site, photographed from the window of the crew cabin. Numerous small rocks and craters can be seen between the LM and the lunar horizon. (NASA photo 69-H-1043.)

of the EVA, to assure that some lunar material would be returned in the event of premature termination of the mission. About halfway through the EVA the bulk sample was collected by filling one of the sample-return containers (rock boxes in popular jargon) with scoops of soil and randomly selected rock fragments. The bulk sample weighed 14.6 kg and consisted of 4.2 kg of rock fragments larger than 1 cm and 10.2 kg of finer material. The documented sample, collected at the end of the EVA, comprised about 20 selected rock fragments weighing 6.0 kg; this sample was sealed in a second rock box along with two filled core tubes and the aluminum foil that had been deployed to investigate the solar wind. The total weight of lunar material returned was 22 kg. (The major limitation on the return of lunar material is the storage capacity and the weight-lifting capability of the ascent stage of the LM.)

After splashdown in the Pacific on July 24 the rock boxes were flown to the Lunar Receiving Laboratory (LRL) at the Manned Spacecraft Center at Houston, arriving there July 25. The boxes were put into a decontamination chamber, where terrestrial contamination was removed by a peracetic acid spray; after rinsing and drying, the boxes were passed into a complex vacuum chamber equipped with spacesuit arms good for 1 atm of differential pressure. Here the boxes were opened and the samples processed. It had been decided to quarantine the lunar material for 1 month or more, in order to make sure that it was safe to release it. This was to guard against the possibility (however remote) of introducing pathogenic organisms from the Moon. During the quarantine period, however, the samples were intensively investigated by the Lunar Sample Preliminary Examination Team (LSPET), all handling being done in gloved cabinets, either in a vacuum or in a dry nitrogen atmosphere. The rocks were examined, described, photographed, weighed, and chipped for physical and chemical analysis and biological testing. Portions of the returned material were used for the determination of density, reaction to gases, mineralogical composition, chemical analysis by optical emission spectroscopy, and total organic carbon analysis. Rare gases were extracted and analyzed, and counting of short-lived isotopes was performed in a specially shielded underground laboratory. The results of these studies were published in *Science,* September 18, 1969, at the time when the Apollo 11 samples were released from quarantine for distribution to 140 principal investigators around the world.

The LSPET team studying the mineralogy and petrology of the returned materials divided them into four groups, as follows:

Type A: Fine-grained vesicular crystalline igneous rocks, analogous to terrestrial basalt (Fig. 2-4).

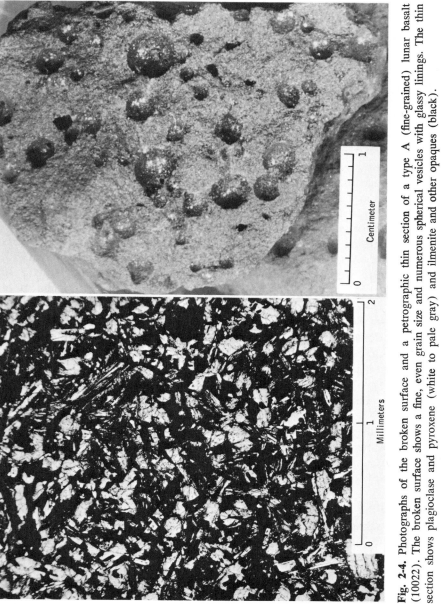

Fig. 2-4. Photographs of the broken surface and a petrographic thin section of a type A (fine-grained) lunar basalt (10022). The broken surface shows a fine, even grain size and numerous spherical vesicles with glassy linings. The thin section shows plagioclase and pyroxene (white to pale gray) and ilmenite and other opaques (black).

Type B: Medium-grained vuggy crystalline igneous rocks, analogous to terrestrial dolerite (Fig. 2-5).

Type C: Microbreccias, consisting of small rock and mineral fragments in a groundmass of comminuted material (Fig. 2-6).

Type D: Fines, material smaller than 1 cm in diameter.

The LSPET team numbered the Apollo 11 specimens in a series beginning with 10001; this series runs to 10085, and publications describing these specimens may use the full number or just the last two digits.

Two types of unique surface features occur on all rocks of the lunar collections: small pits lined with glass, and glass spatters not necessarily associated with pits. The diameters of the pits average 1 mm or less. The glass surfaces in the pits are brightly reflecting and are commonly uneven and botryoidal. Presumably, the pits have been caused by the impact of small particles traveling at very high velocities. The glass spatters may be related to nearby impact events.

A conference to report the results obtained by the principal investigators of the Apollo 11 samples was held in Houston on January 5–8, 1970. These results were recorded in the form of brief articles in the January 30 issue of *Science,* and in more extended articles in a 1970 supplement of *Geochimica et Cosmochimica Acta.*

The Apollo 12 mission took off on November 16, 1969. The LM touched down in the eastern part of Oceanus Procellarum within 200 m of the Surveyor III spacecraft, which had landed there over two years earlier, in April 1967 (Fig. 2-7). The landing site (Fig. 2-8) is characterized by a distinctive cluster of craters ranging from 50–400 m in diameter. The informal names applied to these craters are indicated on Fig. 2-8. The extravehicular activity on this mission was divided into two 4-hr periods separated by an 8-hr eat-rest period. Four separate samples were collected: a contingency sample (1.9 kg), a selected sample (14.8 kg), a documented sample (11.1 kg), and a tote-bag sample (6.5 kg), a total sample return of 34.3 kg. Beside collecting samples, the chief scientific tasks during the EVA were (1) photographing rocks, the lunar surface, and the moonscape; (2) deploying the Apollo Lunar Surface Experiments Package (ALSEP); (3) collecting parts from the Surveyor III spacecraft; and (4) deploying and returning the solar wind foil. The ALSEP contained a central power station and five experimental units: (1) a passive seismometer; (2) a cold-cathode ion gage; (3) a supra-thermal ion detector; (4) a solar-wind spectrometer; and (5) a magnetometer.

On return to the LRL the Apollo 12 samples were handled in a similar fashion to those of Apollo 11, being placed in quarantine and examined

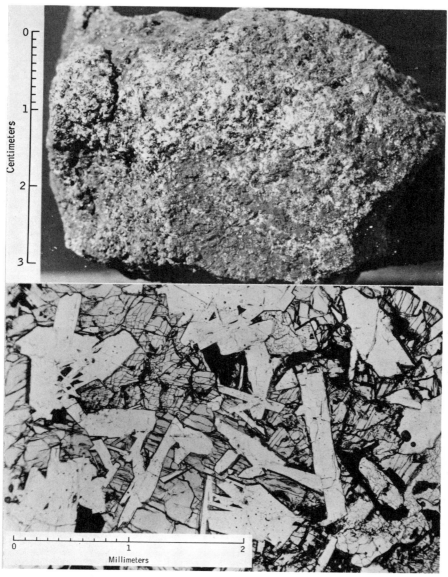

Fig. 2-5. Photographs of the broken surface and a petrographic thin section of a type B (medium-grained) lunar basalt (10047). The broken surface shows a holocrystalline granular texture with comparatively large grain size. The thin section shows a microgabbroic texture with laths of plagioclase (white), together with pyroxene (gray) and minor amounts of ilmenite and other opaques (black).

Fig. 2-6. Photographs of the broken surface and a petrographic thin section of a type C microbreccia (10019). The surface shows rock and mineral fragments in a fine-grained matrix, with glass-lined pits and glass spatters. The thin section shows the comminuted rock and mineral fragments in a semiopaque, glass-rich matrix.

Fig. 2-7. The Surveyor III spacecraft (foreground) and the Apollo 12 LM (on the lunar horizon), a photograph taken during the second EVA of the mission; the distance between the two is about 180 m. (NASA photo 69-H-1989.)

in a closed system in a vacuum or an atmosphere of dry nitrogen. The LSPET team found the same rock types as in the Apollo 11 collections, but the proportions of the types were different (LSPET, 1970). At the Apollo 12 site the rocks are predominantly crystalline, whereas at the Apollo 11 site the rocks were about half crystalline and half microbreccia. The LSPET team speculates that this difference is probably due to the fact that the rocks from the Apollo 12 mission were collected primarily on or near crater rims. On crater rims the regolith is thin or only weakly developed, and many of the rocks collected are probably derived from crater bedrock. The Apollo 11 site is on a thick mature regolith, where

24   THE LUNAR ROCKS

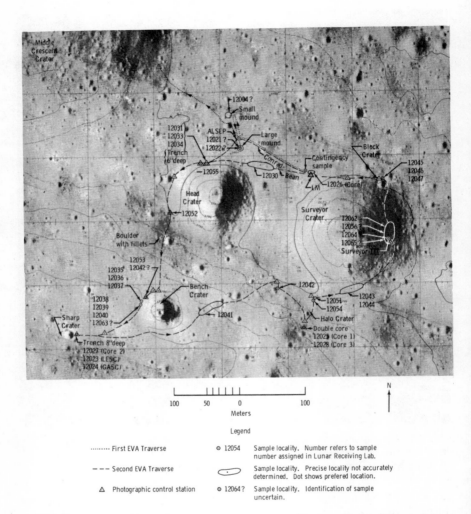

Fig. 2-8. Traverse map of the Apollo 12 site, showing sample localities and topographic features.

many of the rock fragments have been produced by the lithification of the regolith itself.

Of the rocks collected on the Apollo 12 mission, four weighed more than 2 kg each, six between 1 and 2 kg, and seven between 500 and 1000 g. The different samples have been numbered in a series beginning with

12001 and running through 12077. The same division into types A, B, C, and D as was devised for the Apollo 11 collections is employed in the preliminary description of the Apollo 12 materials. Most of the rocks are described as basalts and olivine dolerites. Their mineralogy appears to be essentially the same as for the Apollo 11 rocks, the principal difference being in the lower abundance of ilmenite ($FeTiO_3$), and the higher abundance of olivine, $(Mg, Fe)_2SiO_4$. Some of the rocks are notably coarse-grained, almost pegmatitic in appearance.

For the Apollo 13 mission the site selected was the Fra Mauro region, in a highland area to the east of the Apollo 12 site. This region appears to be blanketed by throw-out from the impact that excavated Mare Imbrium, and the rocks should thus be older than those flooring the mare sites visited on Apollo 11 and 12. Other materials in the region have been provisionally mapped as post-Imbrian volcanic rocks. Unfortunately this mission, after a flawless blastoff on April 11, 1970, had to be aborted in midcourse because of a serious malfunction in the service module. The spacecraft orbited the Moon without landing and returned safely to Earth on April 17.

Candidate sites have been selected for future lunar landings (Fig. 2-9). The rationale of site selection is based on the premise of providing a balanced list of scientific sites for lunar exploration for the remaining Apollo flights. In the broadest sense, the aim is to obtain the maximum information bearing on the origin and history of the Moon. More specific reasons for site selection are:

1. To determine the composition and structure of the mare regions, and their variation.
2. To determine the composition and structure of the highlands, and their variation.
3. To determine the composition and structure of deep-seated material, both under the mare and under the highlands.
4. To elucidate the nature of lunar processes such as tectonism, rille formation, volcanism, and impact phenomena.

The Apollo 14 mission, which may be flown in January 1971, was provisionally scheduled for the Littrow area on the southeastern margin of Mare Serenitatis. This is a fairly level cratered plateau, sometimes called the Serenitatis Bench, between a typical mare to the west and rugged highlands to the east. The area is blanketed by very dark material (some of the darkest on the moon's surface), which suggests a distinctive composition; it may be relatively young and of volcanic origin. The area also has a rille network and an abundance of fresh-looking wrinkle ridges.

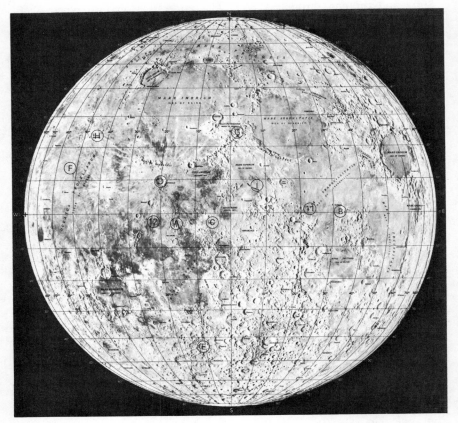

Fig. 2-9. The near side of the Moon, showing present and projected Apollo landing sites: 11 = Apollo 11; 12 = Apollo 12; A = Fra Mauro; B = Censorinus; C = Mösting C; D = Copernicus; E = Tycho; F = Marius Hills; G = Hadley's Rille; H = Aristarchus (Schroeter's Valley); I = Hyginus Rille.

However, it is possible that Apollo 14 may be directed to the Fra Mauro region, which was the target for Apollo 13.

A variety of sites are potential candidates for the remaining missions, tentatively scheduled through Apollo 19, to be flown in 1974. Later missions include more ambitious EVA programs, which may be extended to 6 hr from the present 4 hr, and to three per mission instead of the present two. Apollo 14 and later missions may have a small handcart for carrying equipment and samples on the EVA, and a powered roving vehicle is planned for Apollo 16. Such a vehicle would extend the mobility of the astronauts to several kilometers around the landing site, and could possibly be programmed for a lengthy unmanned traverse across the lunar surface. Hopefully, the amount of lunar material returned to the earth by each mission will also be increased to about 150 kg.

Of the possible sites for later Apollo missions, a prime candidate is Censorinus, a small extremely fresh crater in highland terrain on the southern side of Mare Tranquillitatis, somewhat to the east of the Apollo 11 landing. A landing here can be expected to achieve three main objectives: to establish the age of what is clearly a very young feature on the Moon's surface, to investigate and characterize an unquestioned impact feature, and to obtain samples of highland material. An alternative site offering similar possibilities is Mösting C, a small fresh crater in highlands on the eastern side of Oceanus Procellarum.

Another goal of the later Apollo missions is the exploration of one or more of the very large rayed craters, of which Copernicus and Tycho are the likely candidates. These large craters are of interest not only because they represent major events in the history of the Moon, but also because they probably expose material from a range of depths up to 10 km, or even more. A mission to Tycho would probably aim to land on the rim near the Surveyor VII spacecraft. This would enable the astronauts to collect information and samples from the immediate vicinity of the spacecraft, in order to evaluate further the technological efficiency and the scientific accuracy of the Surveyor instruments for future lunar and planetary missions. The area of exploration would be extended beyond that examined by Surveyor VII. This would include such important features as the large blocks of rock ejected from Tycho itself, the flowlike surface blanket, and the material of the prominent rays. A landing in Copernicus would probably be targeted for the crater floor, with the aim of examining and collecting from both the central peaks and the layered wall materials (Fig. 2-10); to be most effective, such a mission would require a long stay time and the use of a powered roving vehicle.

A completely different type of moonscape would be investigated by a mission to the Marius Hills region, a complex of domes, mare ridges, and sinuous rilles in the western part of Oceanus Procellarum (Fig. 2-11). The setting and structure of this region suggest that it has been an area of volcanic activity. The mare ridges may be surface expressions of convection currents within the Moon, analogous to the midoceanic ridges on Earth. As a volcanic province, the Marius Hills would provide an opportunity to sample a sequence of material extruded from the Moon's interior and then subjected for varying lengths of time to lunar and extralunar processes.

Another prime area for an Apollo landing is the region of the Apennine

28   THE LUNAR ROCKS

Fig. 2-10. Oblique view of the crater Copernicus, 90 km in diameter, showing details of the floor, the central mountains, and the northern wall. Flow lines, ridges, and troughs runnning down the far wall indicate vast landslides off the wall onto the floor. Large blocks litter the slopes of the central mountains, which rise up to 800 m above the floor. (NASA Orbiter II photo.)

Mountains, which rise thousands of feet above the southeastern part of Mare Imbrium. The precise landing site would probably be near Rima Hadley (Hadley's Rille), a sinuous rille running parallel with the Apennine front; a particularly interesting spot would be the vicinity of Hadley C, a small (5.5 km in diameter) sharp crater, whose ejecta appear to have partly covered the rille (Fig. 1-3). A landing here would shed light on the origin of rilles, and provide an opportunity to study and collect from the extensive vertical section of the highland scarp provided by the Apennine front. This mission would be greatly facilitated by the availability of a powered roving vehicle. After use on the mission the vehicle could possibly be sent on a long unmanned traverse from Rima Hadley into Mare Imbrium and thence into Mare Serenitatis. Along the way it could provide continuous profiles of variations in gravity, magnetic and electric fields, and thickness of the surface layer. This particular traverse crosses one of the largest mascon areas; it would cover enough ground to explore the phenomenon adequately with geophysical techniques. The continuous monitoring of gravity would also provide information on the

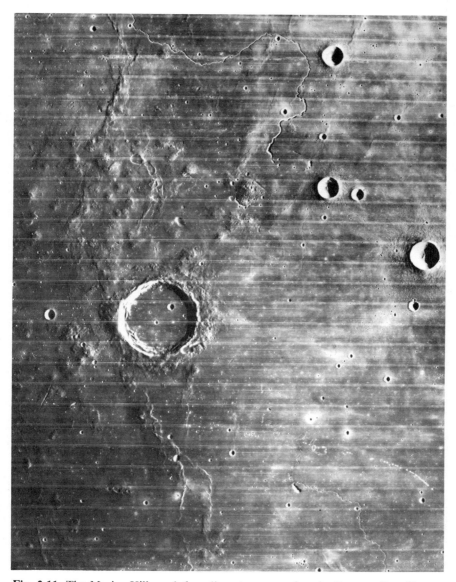

Fig. 2-11. The Marius Hills and the adjacent mare surface in Oceanus Procellarum. The large crater in the center is Marius, 41 km in diameter (note sharp younger craters in the floor). The Marius Hills are characterized by a large number of domelike hills rising several hundred meters; they have been interpreted as shield volcanoes and as laccolithic intrusions. Also noteworthy are the wrinkle ridges and the sinuous rilles. (NASA Orbiter IV photo.)

30   THE LUNAR ROCKS

regional isostatic balance of the Moon, that is, whether the higher topographic features are compensated by a deficiency of mass beneath them or whether they represent loads on the surface.

Another rille of particular interest is that known as Schroeter's Valley, near the large crater Aristarchus to the north of Oceanus Procellarum (Fig. 2-12). Schroeter's Valley is one of the largest of the lunar rilles, and forms a sinuous depression up to 1300 m deep and several kilometers across. It terminates at one end in a large deep depression, called from its shape the Cobra Head, which is interpreted as its source. From the Cobra Head also springs a smaller sinuous rille that meanders along the

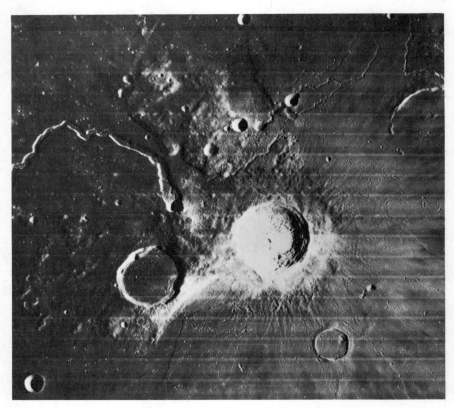

Fig. 2-12. The bright crater Aristarchus (45 km in diameter) is surrounded by an ejecta blanket and secondary impact craters. Superposition of the latter on the surrounding terrain shows that Aristarchus is younger than these materials. The prominent rille is Schroeter's Valley, beginning at the "Cobra Head," in a hill that rises about 1500 m above the adjacent plateau. The large shallow crater near the "Cobra Head" is Herodotus, 35 km in diameter. (NASA Orbiter IV photo.)

whole length of the valley floor. Similar median rilles have been observed in other lunar valleys. The vicinity of Schroeter's Valley also shows numerous cones and domes, which may be of volcanic origin. This suggests that a variety of rock types may be sampled in this region.

Another interesting site would be the Hyginus Rille in Mare Vaporum. This rille apparently branches from the crater Hyginus (10 km in diameter) at two points, extending toward the northwest and east-southeast. It is a linear rille, almost straight, in contrast to the sinuous rilles such as Rima Hadley, and is noteworthy for a number of craters spotted along its length. Its linearity, and the crater chain associated with it, suggest that it may represent a tectonic rift that has triggered volcanism along is length.

CHAPTER 3

LUNAR MINERALOGY

Compared to terrestrial rocks, the mineralogy of the lunar samples so far collected is quite limited. Of course, these samples have been collected from very small areas, although they evidently include material ejected from other areas of the Moon and material introduced by impacting meteorites. The limited mineralogy can also be ascribed in part to the limited range in chemical composition, and in part to the absence of water and weathering processes on the Moon. On the Earth these processes result in the formation of a great variety of hydrated minerals from the comparatively few primary minerals of igneous rocks.

The following table of lunar minerals has been compiled from the numerous reports on the Apollo 11 materials. The division into major, minor, and accessory minerals is convenient but to some extent arbitrary. For example, in some rocks ilmenite and pyroxene are only minor constituents, iron and nickel-iron may slightly exceed 1% in some samples, and the minor and accessory minerals may be found only in specific rock types.

Besides the minerals listed in Table 3-1, a few more have been tentatively identified, but further information is needed before their occurrence as lunar minerals can be accepted as certain. These include aragonite, $CaCO_3$ (Gay et al., 1970); chalcopyrite, $CuFeS_2$ (Agrell et al., 1970); magnetite (Gay et al., 1970); an amphibole mineral (Gay et al., 1970); a mineral of the mica group (Arrhenius et al., 1970); graphite (Arrhenius et al., 1970); and hematite (Ramdohr and El Goresy, 1970).

The report on the preliminary investigation of the Apollo 12 collections (LSPET, 1970) indicates that their mineralogy is similar to those of Apollo 11, although the proportions of the major minerals differ (Fig. 3-1). Plagioclase, pyroxene, olivine, cristobalite, ilmenite, potash feldspar (sanidine), troilite, and metallic iron have been positively identified.

LUNAR MINERALOGY

Table 3-1. Minerals in the Lunar Materials

	Name	Formula	Crystal system
Major (>10%)			
	Pyroxene	$(Ca, Fe, Mg)_2Si_2O_6$	Monoclinic
	Plagioclase	$(Ca, Na)(Al, Si)_4O_8$	Triclinic
	Ilmenite	$FeTiO_3$	Hexagonal
Minor (1–10%)			
	Olivine	$(Mg, Fe)_2SiO_4$	Orthorhombic
	Cristobalite	SiO_2	Isometric (high-temperature)
	Tridymite	SiO_2	Hexagonal (high-temperature)
	Pyroxferroite	$CaFe_6(SiO_3)_7$	Triclinic
Accessory (<1%)			
	Copper	Cu	Isometric
	Iron	Fe	Isometric
	Nickel-iron	(Fe, Ni)	Isometric
	Cohenite	Fe_3C	Orthorhombic
	Schreibersite	$(Fe, Ni)_3P$	Tetragonal
	Troilite	FeS	Hexagonal
	Potash feldspar	$KAlSi_3O_8$	Monoclinic
	Quartz	SiO_2	Hexagonal
	Armalcolite	$(Fe, Mg)Ti_2O_5$	Orthorhombic
	Ulvöspinel	Fe_2TiO_4	Isometric
	Chromite	$FeCr_2O_4$	Isometric
	Spinel	$MgAl_2O_4$	Isometric
	Perovskite	$CaTiO_3$	Isometric
	Rutile	TiO_2	Tetragonal
	Baddeleyite	ZrO_2	Monoclinic
	Zircon	$ZrSiO_4$	Tetragonal
	Apatite	$Ca_5(PO_4)_3(F, Cl)$	Hexagonal
	Whitlockite	$Ca_3(PO_4)_2$	Hexagonal

Spinel, tridymite, metallic copper, and pyroxferroite were tentatively identified by optical means. Olivine is a major mineral in many of the rocks, and ilmenite is a minor, not a major, component.

A discussion of the individual minerals and mineral groups follows.

Pyroxene

Pyroxenes with a wide range of compositions are the principal phases in most of the lunar materials. In the coarser-grained rocks this mineral is readily distinguished with the naked eye as cinnamon brown, lustrous grains up to 2 mm across. In thin sections under the microscope the color

34 THE LUNAR ROCKS

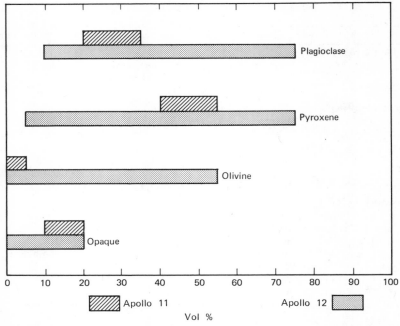

Fig. 3-1. The amounts of the major minerals in the Apollo 11 and Apollo 12 rocks; opaque is largely ilmenite. (LSPET, 1970.)

is usually pale buff, sometimes with a pink tint. Pleochroism is absent or weak. Zoning is common and is best observed under crossed nicols; hourglass structure is sometimes seen. Some grains show simple twinning on (100).

Overlooking for the moment their minor content of titanium and aluminum, we can express lunar pyroxene compositions in terms of the three components $CaSiO_3$ (Wo), $MgSiO_3$ (En), and $FeSiO_3$ (Fs). Many names have been applied to different pyroxene compositions, but the current standard nomenclature is illustrated in Fig. 3-2; the shaded areas in this figure indicate compositions for which natural pyroxenes are unknown. (The composition of the new mineral pyroxferroite falls within the shaded area near the $FeSiO_3$ apex.)

Individual compositions of lunar pyroxenes are also shown in Fig. 3-2. Most of them fall within the fields of augite and ferroaugite; some are pigeonites, frequently as cores mantled by augite. Exsolution of hypersthene or clinohypersthene is rarely visible in microscopic examination but is frequently detected by x-ray diffraction or electron microscopy. The range in compositions illustrated in Fig. 3-2 may be observed in a single

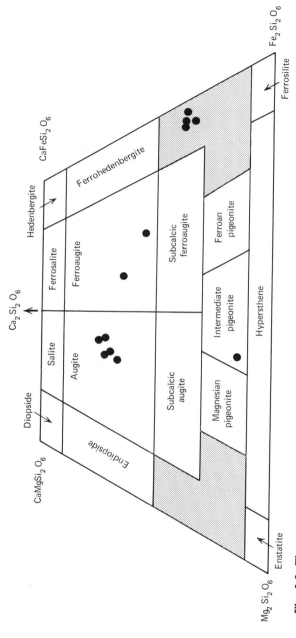

Fig. 3-2. The pyroxene system, showing augite and ferroaugite compositions in lunar crystalline rock 10047; and the composition of the yellow pyroxenoid mineral pyroxferroite in the shaded field, a compositional region for which natural pyroxenes were previously unknown. The composition of an intermediate pigeonite grain in crystalline rock 10003 is also shown. (Mason et al., 1970.)

crystal, zoned from a magnesium-rich core to an iron-rich margin. Sometimes the pyroxene is rimmed with pyroxferroite, indicating that when the Fe/Mg ratio exceeds a certain value a structural change is induced, with the formation of this new mineral.

The lunar augite and ferroaugite contain appreciable amounts of titanium and aluminum, amounts up to about 5% TiO_2 and 6% Al_2O_3 having been recorded. Yagi and Onuma (1967) have shown that titanium in augite can be considered as incorporated in the form of a hypothetical titan-pyroxene component, $CaTiAl_2O_6$. They found that the maximum solubility of this component in $CaMgSi_2O_6$ is about 11% at atmospheric pressure, corresponding to about 4% TiO_2 by weight, and that solubility decreased with increasing pressure. However, the solubility of $CaTiAl_2O_6$ seems to be higher than this in the lunar pyroxenes, to account for the high titanium content in some of them.

Manganese and chromium are present in the lunar augite in small amounts (Cr $\sim 0.3\%$, Mn $\sim 0.2\%$). Sodium, commonly present in terrestrial augites, is absent or present in only trace amounts ($<0.1\%$). Pigeonite contains lower concentrations of trace elements than augite (Cr $\sim 0.2\%$, titanium up to 1%, aluminum up to 0.5%), except for manganese ($\sim 0.5\%$).

Orthopyroxene is essentially absent in optically recognizable grains in the type A and B crystalline rocks. It is, however, frequently found in the microbreccias and regolith samples. Fragments of strongly crushed plagioclase-orthopyroxene rocks, orthopyroxene with broad kink bands, and single pyroxene grains showing ortho- and clinopyroxene exsolution have been reported. Such pyroxenes most likely are not fragments of the type A and B samples, and both distant lunar sources and meteorite increments are postulated sources. The well-developed exsolution texture noted by Fredriksson et al. (1970) clearly required much slower cooling conditions than prevailed during cooling of even the type B samples, in which exsolution is mainly submicroscopic.

Pyroxferroite

The LSPET team observed a yellow mineral in the coarser-grained Apollo 11 rocks, which they recognized to be probably a new mineral. They noted that it seemed to be concentrated in vuggy areas of the rocks, evidently a late crystallization. This mineral was subsequently examined by several groups of investigators and proved to be structurally identical with the terrestrial mineral pyroxmangite. Pyroxmangite is essentially a manganese silicate in which some of the manganese is replaced by ferrous ion.

The lunar mineral is essentially a ferrous silicate with only minor amounts of manganese, so that it merits a specific name; *pyroxferroite* was selected as indicating its composition and its relationship to pyroxmangite.

The mineral occurs mostly as anhedral grains, occasionally as euhedral crystals, up to a few tenths of a millimeter in size; it has been observed mantling a core of pyroxene. Individual grains are yellow, but the mineral is almost colorless in thin section. It is triclinic; the cell dimensions are $a = 6.62$, $b = 7.55$, $c = 17.35$ Å; $\alpha = 114.5°$, $\beta = 82.7°$, $\gamma = 94.5°$. Measured specific gravities are 3.68 and 3.76; calculated from cell dimensions and chemical composition, 3.83. Pyroxferroite is optically positive, with refractive indices $\alpha = 1.750$–1.756, $\beta = 1.752$–1.758, $\gamma = 1.766$–1.768; $2V = 35$–$40°$. It has perfect (110) and (1$\bar{1}$0) and poorer (100) cleavages.

Many analyses have been made; they show a somewhat variable composition, as follows (wt %): SiO_2, 44.7–47.1; TiO_2, 0.2–0.7; Al_2O_3, 0–1.2; FeO, 44.6–47.7; MgO, 0.3–1.2; MnO, 0.6–1.0; CaO, 4.7–6.3; Na_2O, 0–0.1; K_2O, trace. In terms of the principal components the formula can be written $CaFe_6(SiO_3)_7$, which would be the content of one-half the unit cell. This formula is consistent with the crystal structure, the principal feature of which is a single chain of SiO_4 groups linked through common oxygen atoms (giving the SiO_3 composition), the chain having a 7-unit repeat. The pyroxene chain has a 2-unit repeat; minerals having SiO_3 chains but with different repeats are known as pyroxenoids. Pyroxferroite can thus be classed as a pyroxenoid.

As mentioned previously, the composition of pyroxferroite places it in an area of the pyroxene quadrilateral (Fig. 3-2) where natural pyroxenes are unknown. Such compositions are absent or very rare in terrestrial rocks, since they require high iron contents and highly reducing conditions; in laboratory syntheses they usually crystallize as fayalite (Fe_2SiO_4) and free silica. It is of some interest that a complex calcium-iron-manganese-magnesium silicate [approximately $Ca_2Fe_2Mn_2Mg(SiO_3)_7$], described from steel-furnace slags many years ago and named vogtite (Hallimond, 1919), may be related to pyroxmangite and pyroxferroite.

Plagioclase

After pyroxene, plagioclase feldspar is the most abundant mineral in the Apollo 11 rocks, making up 20–40% by volume in most of them, and somewhat less by weight. Some rock fragments in the lunar breccias and soil consist almost entirely of plagioclase. In the coarser-grained rocks it is seen as white granular aggregates or as platy crystals tabular parallel to (010); individual crystals range up to about 1 mm long (Fig. 3-3).

In thin sections the mineral is colorless and usually shows albite twinning; the crystals are sometimes hollow, a feature characteristic of relatively rapid growth; and the tubular cavities are filled with a fine-grained mesostasis, which evidently represents the last liquid to crystallize.

Plagioclase can be considered a solid solution of the components albite, $NaAlSi_3O_8$ (Ab) and anorthite, $CaAl_2Si_2O_8$ (An). Individual plagioclase grains in the lunar rocks show a range of composition from about An_{60} to nearly pure anorthite (An_{100}), with the commonest composition near An_{90} (Fig. 3-4). Thus most of the lunar plagioclases fall within the

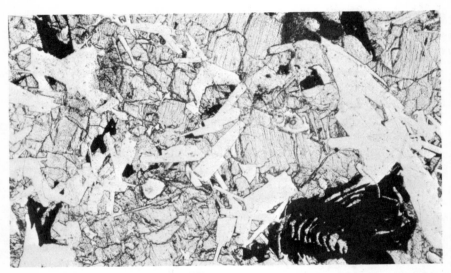

Fig. 3-3. A coarsely crystalline lunar rock (10047), consisting of ilmenite (black), pyroxene (gray), and plagioclase (white). The maximum length of the plagioclase laths is about 1 mm.

composition range of anorthite (An_{90-100}) and bytownite (An_{70-90}); a few fall in the range of labradorite (An_{50-70}). Individual plagioclase crystals may be zoned, becoming more sodic from core to margin.

Beside the major elements, the plagioclase contains minor amounts of potassium ($\sim 0.1\%$), iron ($\sim 0.5\%$), and titanium ($\sim 0.1\%$). The content of titanium is noteworthy, since terrestrial plagioclases usually contain less than 0.01% titanium (Corlett and Ribbe, 1967). The high titanium in the lunar plagioclase is presumably due to the high concentration of this element in the parent magma, and perhaps the extremely reducing environment.

In the lunar breccias and soils the plagioclase grains frequently have a chalky appearance and are very friable. In thin sections under the microscope these grains no longer show the characteristic birefringence of plagioclase—some have become completely isotropic and amorphous to x-rays (Fig. 3-5). Isotropized plagioclase was first observed in the Shergotty meteorite many years ago and was named maskelynite. It is now known that this isotropization is produced by extreme shock; maskelynite has been identified in rocks associated with terrestrial meteorite craters and in those exposed to nuclear explosions. The presence of maskelynite

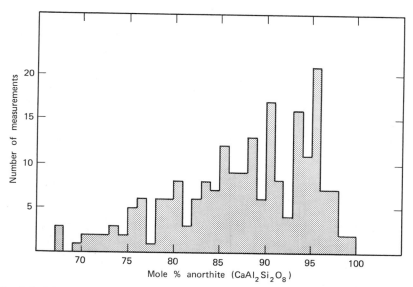

Fig. 3-4. Histogram showing the distribution of anorthite content in 220 grains of plagioclase from the Apollo 11 rocks; three grains not plotted had compositions An_{65}, An_{64}, and An_{58}.

in lunar breccias and soils is therefore good evidence for shock metamorphism related to meteorite impacts on the lunar surface.

Potash Feldspar

Potash feldspar has been recognized as an accessory constituent in the lunar samples. Potassium is a minor constituent of the lunar rocks (0.1–0.2%), and some of this element is combined in the plagioclase feldspar. Part of it, however, evidently concentrated in the residual melt after most of the plagioclase, pyroxene, and ilmenite had formed, and crystallized as

40 THE LUNAR ROCKS

Fig. 3-5. Normal (top) and shocked (bottom) basalts from the Apollo 11 collections. In the shocked fragment the plagioclase has been converted to isotropic maskelynite (black), whereas in the unshocked rock it is birefringent (white and gray), magnification 80×.

small grains of potash feldspar in the mesostasis. It probably contains much of the rubidium, cesium, and barium present in the source magma. The potash feldspar in the Apollo 12 rocks has been identified as the monoclinic variety known as sanidine.

Ilmenite

One of the most remarkable features of the Apollo 11 materials is the abundance of ilmenite. Ilmenite, while common in terrestrial basalts, seldom exceeds 5%, whereas some of the lunar rocks contain 10–20% of this mineral. In the coarser-grained rocks ilmenite can be recognized with the naked eye as black lustrous grains or platy crystals up to 2 mm long.

Chemical analyses and Mössbauer spectra show that the composition of the lunar ilmenite is close to the ideal formula $FeTiO_3$; unlike terrestrial ilmenite, it contains no detectable ferric iron. Some magnesium (up to about 6% MgO) substitutes for ferrous iron; the following minor and trace elements have also been recorded: Al_2O_3, 0.1–0.3; MnO, 0.3–0.6; CaO, 0–0.3; Cr_2O_3, 0.1–1.3; V_2O_3, < 0.1–0.2. The lunar ilmenite is evidently the host for some of the zirconium recorded in the chemical analyses. Arrhenius et al. (1970) record 0.03–0.3% zirconium in this mineral, with large variations from grain to grain and within single grains; they comment that extensive and variable zirconium substitution in ilmenite is consistent with a high temperature of formation and rapid cooling.

Olivine

This mineral is a minor constituent (up to 5%) in some of the Apollo 11 rocks and a major constituent of many of the Apollo 12 rocks. In thin sections it is seen as colorless grains usually mantled by augite (Fig. 3-5). The composition is usually around Fa_{30} [i.e., 30 mole % of the Fe_2SiO_4 (Fa) component] but may range from about Fa_{20} to Fa_{50}. In comparison to most terrestrial olivine it has notably high calcium (0.2–0.4%) and chromium (0.1%), and contains very little nickel (up to about 0.01%); other minor elements are manganese (0.2–0.3%), titanium (0.1–0.2%), and aluminum (about 0.05%).

Another variety of olivine has been observed in trace amounts in the mesostasis of some of the lunar crystalline rocks. This is the extremely iron-rich form, fayalite, close to Fe_2SiO_4 in composition.

Cristobalite, Tridymite, and Quartz

These three polymorphs of SiO_2 have all been recognized in the lunar materials. Cristobalite and tridymite are minor constituents (up to 5%) of some of the lunar basalts, whereas quartz is rare. In thin sections cristobalite is easily recognized as a colorless mineral with very low birefringence, occurring as granular aggregates interstitial to the major minerals; it is evidently one of the last minerals to crystallize. Tridymite is

42 THE LUNAR ROCKS

associated with the cristobalite and occurs as thin platy crystals tabular on (0001).

Naturally occurring cristobalite and tridymite are seldom if ever pure SiO_2, since their structures can accommodate small amounts of other elements, especially alkalies and aluminum; quartz, on the other hand, usually contains only minute traces of other elements. Brown et al. (1970) report 0.4–0.8% aluminum, 0.1% iron, 0.1–0.2% calcium, 0.1% sodium, and 0.2–0.3% titanium in lunar cristobalite.

The presence of tridymite and cristobalite in the lunar rocks provides important evidence as to the temperatures and pressures of crystallization. Tridymite and cristobalite are low-pressure, high-temperature polymorphs of SiO_2 (Fig. 3-6), the pressure stability limit being 3 kb for tridymite

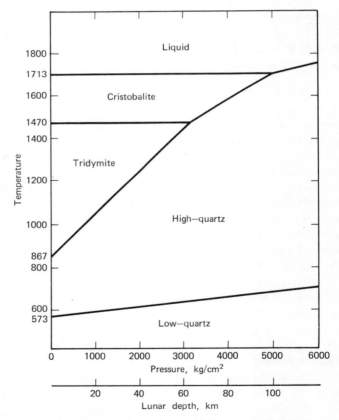

Fig. 3-6. Stability relations of different forms of SiO_2, and the depth-pressure correlation on the Moon.

and 5 kb for cristobalite. These pressures correspond to depths of 60 km and 100 km within the Moon; hence the Apollo 11 rocks cannot have crystallized at great depths. Since the lunar basalts are completely liquid at 1200°C, one might expect tridymite rather than cristobalite in these rocks. However, it is known from terrestrial rocks that cristobalite readily forms at temperatures well below its stability field, a feature probably related to the ease of nucleation or perhaps conditioned by its content of minor elements.

Troilite

Troilite is the only sulfide so far positively identified in the lunar samples. It is universally present in small amounts (up to about 0.7%) and usually occurs as small rounded blebs up to 3 mm across. These blebs probably originated as immiscible droplets in the lunar magma; they frequently contain inclusions of metallic iron. The composition of troilite is close to pure FeS; minor amounts of titanium (0.3%), manganese (0.1%), chromium (0.03%), and nickel (0.02%) have been reported.

Evans (1970) found euhedral crystals of troilite implanted on pyroxene, lining a vug in a coarse-grained lunar rock (10050). These crystals were combinations of hexagonal prisms, pyramids, and base. He determined the unit cell dimensions as $a = 5.962$, $c = 11.750$ Å, with a cell content of 12 FeS and a calculated specific gravity of 4.841.

Iron, Nickel-Iron

Metallic iron is present in the lunar materials in at least two distinct forms: one inherent to the lunar rocks and one introduced by impacting meteorites. In the crystalline rocks it occurs within the troilite blebs, usually as rounded inclusions but occasionally showing cubic crystal forms (Fig 3-7). The proportion of metal to sulfide, about 1:6, is close to the composition of the Fe-FeS eutectic, and the association indicates that the metal and troilite separated together as immiscible droplets from the silicate magma. This metal is essentially pure iron, with very little nickel (0.1% or less) in solid solution; Ramdohr and El Goresy (1970) report that it contains up to 0.7% cobalt. The breccias and the lunar soil contain a nickeliferous iron, usually as irregular fragments up to a few millimeters across but sometimes as spherical or lensoid droplets (Fig. 3-8). This metal may or may not be accompanied by troilite and may contain up to 30% nickel and 1% cobalt. The composition of this metal, and the fact that it is confined to the breccias and soil, indicate that it is extralunar and meteoritic in origin.

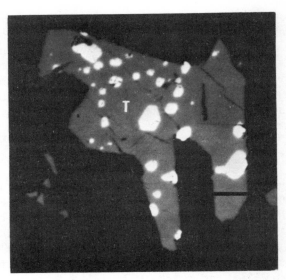

Fig. 3-7. A high-contrast photomicrograph showing spherical blebs and cubo-octahedral crystals of metallic iron (white) in a troilite host (dark gray); scale bar = 0.01 mm. (Haggerty et al., 1970.)

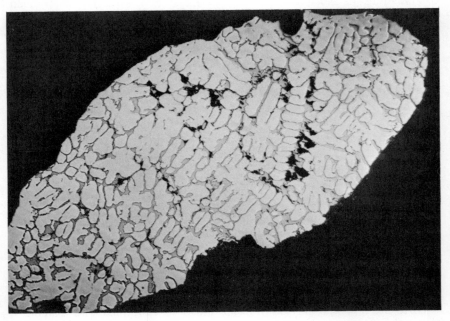

Fig. 3-8. Polished section of a lensoid nickel-iron pellet from lunar fines, showing dendritic aggregate of α_2-metal (white) and troilite (gray); length of section is 2.5 mm. (Mason et al., 1970.)

A probable third form of metallic iron is almost universally present in the lunar glasses as minute particles and thin films (Fig. 3-9). The mode of occurrence of this metal suggests that it may have formed by the direct dissociation of iron-bearing compounds at the high temperatures of glass formation.

The system Fe-Ni has been extensively studied by metallurgists and meteoriticists (Fig. 3-10). Above 910°C nickel-iron alloys exist as the γ-phase or taenite, with a face-centered cubic structure. On cooling, pure iron changes at 910°C into the γ-phase (kamacite), with a body-centered cubic structure. The addition of nickel lowers the temperature of the $\gamma \rightarrow \alpha$ transformation and introduces a two-phase region into the equilibrium diagram, within which the low-nickel α-phase coexists with the γ-phase richer in nickel. However, if cooling is very rapid, equilibrium may not be attained and the γ-phase then inverts into what is known as the α_2-phase.

In the lunar materials the metal in the crystalline rocks and the low-nickel metal in the breccias and soils is present as kamacite, the α-phase, and high-nickel metal (above about 5% nickel) as the α_2-phase (evidently because of very rapid cooling). True taenite may also be present if nickel contents above about 30% are reached.

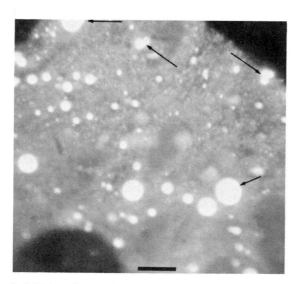

Fig. 3-9. Spherical blebs of metallic iron (white) in a glass fragment from lunar microbreccia 10061; scale bar = 0.01 mm. (Haggerty et al., 1970.)

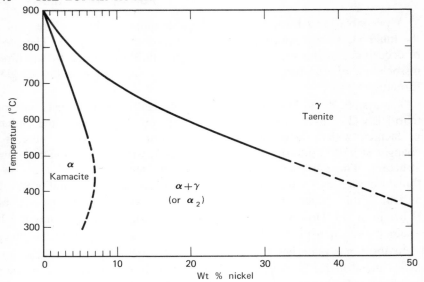

Fig. 3-10. Subsolidus phase relations in the iron-nickel system. (After Goldstein and Ogilvie, 1965.)

Cohenite, Schreibersite

These two minerals have been recognized in trace amounts associated with nickel-iron in breccias and soils (Frondel et al., 1970; Quaide et al., 1970) and are presumably of meteoritic origin. Trace amounts of cohenite or another iron carbide have been identified in the metal-troilite blebs of the crystalline rocks (L. S. Walter, personal communication).

Copper

Native copper, as rare microscopic grains, has been recorded in the Apollo 11 basalts by Ramdohr and El Goresy (1970), and Simpson and Bowie (1970). The latter describe it as occurring in a type A rock (10045) in association with troilite and metallic iron. Copper has also been tentatively identified in the Apollo 12 rocks by the LSPET team (1970). In view of the very low copper content of the lunar rocks (about 10 ppm), free copper or other copper minerals must be extremely rare.

Armalcolite

This mineral was so named to honor the three Apollo 11 astronauts, Neil Armstrong, Edwin Aldrin, and Michael Collins. It is isostructural with the terrestrial mineral pseudobrookite. Pseudobrookite usually occurs as minute crystals in cavities in volcanic rocks; its composition is ferric titanium oxide, Fe_2TiO_5. Armalcolite is a ferrous titanium oxide with considerable magnesium substituting for iron; its formula is $(Fe, Mg)Ti_2O_5$. Electron microprobe analyses give (wt %): TiO_2, 70.3–75.4; FeO, 11.3–18.0; MgO, 5.5–11.1; Al_2O_3, 1.0–2.2; Cr_2O_3, 1.0–2.2; manganese, calcium, and vanadium have been detected in minor amounts (0.1–0.5). Armalcolite is orthorhombic, with $a = 9.743$, $b = 10.024$, $c = 3.738$ Å; the calculated density for $(Mg_{0.5}Fe_{0.5})Ti_2O_5$ is 4.64 g/cm³.

The mineral occurs in the lunar crystalline rocks as minute opaque grains (up to 0.3 mm across), usually as cores to crystals of ilmenite (Fig. 3-11). This indicates that armalcolite probably crystallized early from the lunar magma and its crystallization ceased after a short time, being followed by that of ilmenite.

The relationship of armalcolite to the other iron-titanium oxides in the lunar rocks and to analogous minerals in terrestrial basalts is illustrated in Fig. 3-12. The compositions of the lunar minerals all fall on the $FeO-TiO_2$ join, that is, they contain no ferric iron. The corresponding minerals in terrestrial basalts, whose compositions are indicated by shaded bars in the diagram, always contain appreciable amounts of ferric iron.

Ulvöspinel

This mineral has been recognized as a rare accessory in the lunar material. It occurs as tiny grains and may be intergrown with ilmenite, possibly as a replacement. Keil et al. (1970) give the following analysis (wt %): FeO,

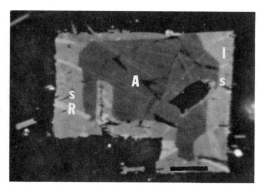

Fig. 3-11. Armalcolite (A) mantled by ilmenite (I). Lamellae of rutile (R) and rods of spinel (S) occur in the ilmenite but terminate at the contact with the armalcolite; scale bar is 0.01 mm. (Haggerty et al., 1970.)

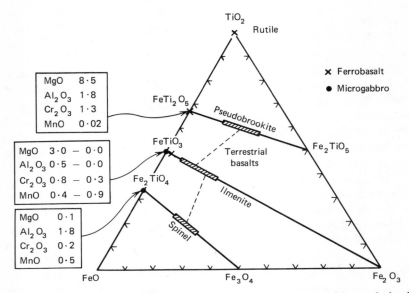

Fig. 3-12. The FeO-Fe_2O_3-TiO_2 system, illustrating the compositions of the lunar minerals, armalcolite, ilmenite, and ulvospinel; comparing them with analogous minerals (shaded bars) in terrestrial basalts. (Anderson et al., 1970.)

62.2; TiO_2, 32.7; Cr_2O_3, 3.5; Al_2O_3, 1.88; MnO, 0.22; MgO, 0.04; and CaO, 0.03. This composition is very close to the ideal formula Fe_2TiO_4, unlike the terrestrial specimens of this mineral, which usually have considerable amounts of Fe_3O_4 in solid solution.

As well as this nearly pure Fe_2TiO_4, another spinel intermediate in composition between this and $FeCr_2O_4$ (chromite) has been recognized. Haggerty et al. (1970) describe it as occurring in a lunar basalt (10020) as euhedral grains 0.1–0.2 mm across, forming about 10% by volume of the iron-titanium oxides in the rock. Many of the grains are mantled by ilmenite. They give the following analysis (wt %): FeO, 42.1; Cr_2O_3, 23.5; TiO_2, 20.9; Al_2O_3, 8.61; MgO, 4.23; V_2O_3, 0.4; MnO, 0.25; and CaO, 0.03. This composition is unknown in terrestrial rocks; chromites are usually low in titanium, and titanium-rich spinels contain little or no chromium.

Chromite

Chromite has been recognized as an inclusion in a nickel-iron pellet and is thus presumably of meteoritic origin (Mason et al., 1970). Grains of chromite, also possibly of meteoritic origin, have been identified in the lunar fines and breccias.

Spinel

Keil et al. (1970) have found nearly pure $MgAl_2O_4$ in a lithic fragment in a lunar breccia. They give the following microprobe analysis (wt %): Al_2O_3, 68.0; MgO, 25.9; FeO, 3.33; and Cr_2O_3, 2.16.

Perovskite

A rare-earth-bearing variety of this mineral, known as dysanalyte, has been recognized in the mesostasis of some of the coarsely crystalline lunar rocks In thin sections it is deep red in color, isotropic, and with high relief. Ramdohr and El Goresy (1970) report that microprobe analyses indicate major calcium, titanium, and iron, with high concentrations of rare-earth elements, zirconium, yttrium, and hafnium, and traces of niobium, barium, and sodium.

Perovskite occurs as an accessory in a few terrestrial igneous rocks, but these are generally of unusual composition, with high Mg/Fe ratios and low in silicon (i.e., ultrabasic, and enriched in titanium). The formation of perovskite as a late crystallization in the lunar rocks probably indicates that nearly all the iron had been removed in pyroxene and ilmenite, thus promoting the formation of a calcium titanium oxide.

Rutile

This mineral has been recognized as minute inclusions in ilmenite, either produced by exsolution or possibly by local reduction ($FeTiO_3 =$ Fe $+$ $TiO_2 +$ O).

Baddeleyite

This zirconium mineral has been identified as small grains in the mesostasis of some of the lunar crystalline rocks and in high-density concentrates from the breccias and fines. Ramdohr and El Goresy (1970) remark that the mineral is rich in hafnium (about 2%) and contains other minor elements such as the rare earths; it occurs as thick tabular untwinned grains, quite different from terrestrial occurrences.

Zircon

This mineral appears to be rarer than baddeleyite in the lunar materials. Dr. U. B. Marvin (personal communication) has found very small amounts in high-density fractions of the lunar fines, as colorless grains, somewhat cloudy with microfractures, uniaxial or slightly biaxial, with refractive indices $\omega = 1.869$, $\varepsilon = 1.901$. Terrestrial zircon is frequently metamict (i.e., the crystal structure has broken down, probably caused by the disintegration of uranium and thorium in solid solution). The lunar zircon has not undergone this breakdown, which suggests that its content of radioactive elements may be very low.

Apatite

As in terrestrial basalts, apatite is the usual phosphate mineral in the lunar rocks. It is an accessory mineral present in amounts of 0.2–0.4% in the crystalline rocks, occurring as small colorless prisms up to about 0.2 mm long in the mesostasis. Apatite apparently varies somewhat in composition, especially in the relative amounts of fluorine and chlorine: Albee et al. (1970) report about 3.4% fluorine in two analyses and about 2.4% fluorine and 1.2% chlorine in another; Keil et al. (1970) state that they identified chlorapatite in grains too small for accurate microprobe analyses. Albee et al. report about 0.6% yttrium and rare earths in apatite lunar.

Whitlockite

This mineral, which is a common accessory in meteorites but has not been recorded in terrestrial basalts, has been identified in lunar rocks by Albee et al. (1970). They record about 9% yttrium and rare earths in this mineral.

Summing up, it may be said that the lunar mineralogy as now known, while not extensive, is extremely interesting. It is clearly analogous to that of terrestrial basalts but reflects an extension into chemical compositions and physicochemical conditions of crystallization unknown in terrestrial rocks. In particular, the low fugacity of oxygen during crystallization of the lunar rocks results in the appearance of free iron and the practical absence of ferric iron, a situation extremely rare on Earth. As a consequence, we find such exotic minerals as troilite, armalcolite, and pyroxferroite. On the other hand, the similarity in the major element chemistry between lunar basalts and their terrestrial equivalents is reflected in the similarity of their bulk mineralogy, dominated by the presence of pyroxene, plagioclase, olivine, and ilmenite.

CHAPTER 4

LUNAR PETROLOGY: THE IGNEOUS ROCKS

INTRODUCTION

Students of the Moon have long wondered about the role of volcanism in lunar history. Were there volcanic rocks? If so, were the melts that produced them generated by large impacting meteorites or by internal processes like those which generate terrestrial lavas? What were their chemical and mineralogical compositions? How would they be distributed in regard to craters, highlands, and maria, and to even more enigmatic surface features such as lunar rilles? What would be their ages? Would they be like terrestrial volcanic rocks? On Earth, volcanic rocks had provided clues about the composition and mineralogy of the upper mantle. Would lunar volcanic rocks be equally informative about the lunar interior? Not surprisingly, then, the astronauts' televised reports of the observing and collecting of probable basaltic volcanic rocks created excitement among lunar scientists.

The preliminary examination team recognized three distinctive groups of lunar samples: the crystalline igneous rocks, the microbreccias, and the fines. This first category is the subject of this chapter. The second and third categories, the subject of the following chapter, includes lunar material modified by meteorite impact and extralunar (meteoritic) increments.

This chapter focuses on the properties of igneous rocks of the Apollo 11 mission. Where data are available, comparisons are made with the recently obtained Apollo 12 samples.

HAND-SPECIMEN FEATURES

A number of important features are readily visible with the unaided eye and with a binocular microscope. The igneous rocks were recognized as such and immediately divided into two classes based on their grain size: type A (fine-grained) (Fig. 2-4) and type B (medium to coarse-grained) (Fig. 2-5). These rocks, which have come to be called lunar basalts, were once totally molten. This is evidenced by their common gas cavities—termed *vesicles* when spherical, and *vugs* when irregular and lined with projecting crystals—and their textures.

The lunar igneous rocks are mainly volcanic rocks, that is, they cooled at, or very near, the lunar surface. This is indicated by the small grain size of many, which is comparable to rapidly cooled basaltic igneous rocks on earth. Indeed, it is difficult to distinguish at a glance the type A samples from certain terrestrial basaltic lavas.

The coarser-grained rocks (type B) certainly did not cool in immediate contact with the vacuum of the lunar surface; their grain size would be much finer, like those of the type A "basalts." Rather, they cooled slowly, insulated by either the already cooled upper surface of a lava lake or were intruded at shallow depth. Rocks of the latter category are termed *hypabyssal;* they are commonly the dikes and sills that underlie volcanic regions and often serve as feeders for the surface eruptions.

Gas cavities (vesicles and vugs) are common in the lunar basalts and reflect the presence of a gas phase that was emitted from the melt during their eruption or shallow intrusion. Such gas emission normally takes place because the pressure on the liquid is decreased—a phenomenon not unlike the bubbling-over of just-opened carbonated beverages. As molten rock, termed *magma,* ascends from the lunar interior, it is under less and less confining pressure. If the dissolved gases exert a pressure greater than the confining pressure, gas begins to exsolve from the magma, forming gas cavities. The composition of the gas that originally exsolved from the magma is unknown. In the type B or coarser-grained lunar basalts, the gas cavities are commonly irregular, reflecting some infiltration of liquid and crystals into the cavities.

MINERALOGICAL COMPOSITION AND PETROGRAPHY

Most Apollo 11 and 12 igneous rocks are dark colored and contain various proportions of the principal minerals: pyroxene, plagioclase, olivine, and ilmenite. The Apollo 11 samples are a relatively homogeneous group, showing only small ranges in relative abundances of minerals. In contrast the Apollo 12 samples range more widely (Fig. 4-1).

There are small variations in composition and larger variations in grain size and texture (Figs. 4-2 and 4-3) in the Apollo 11 crystalline rocks.

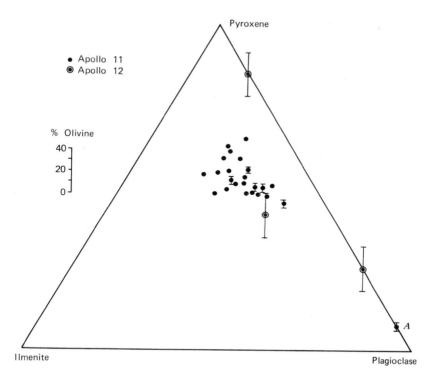

Fig. 4-1. The Apollo 11 igneous rocks show smaller ranges in mineralogical composition than those of Apollo 12. The percent of olivine is indicated by the length of the vertical line through the point (note scale at left). Minor minerals are not shown. The plagioclase-rich sample (lower right, labeled *A*) is a small fragment from the Apollo 11 fines (Chao et al., 1970). Note that most Apollo 12 samples contain considerably more olivine than those of Apollo 11. Apollo 11 modal analyses from a number of unpublished and published sources. Apollo 12 modal analyses are from the LSPET (1970).

The principal minerals are augite, pyroxene, and ilmenite. Augite typically ranges from about 47 to 55 vol % and averages around 50%. The abundance of augite and ilmenite account for the dark color of these rocks. The range in plagioclase and ilmenite contents is larger: 24–37%, and 9–19%, respectively. The average grain size ranges from about a few hundredths of a millimeter, up to slightly over 2 mm. The finer-grained lunar rocks correspond to quenched surfaces of lava flows, or to ejected cinders or bombs, and the coarser-grained samples correspond to flow interiors or perhaps to the interiors of thin sills or dikes.

Fig. 4-2. The grain size and texture ranges widely between different Apollo 11 igneous rocks: fine-grained sample type A (10049) on left, coarse-grained sample type B (10047) on right. The middle sample is intermediate-grain size, also termed type B (10020). The width of the field (0.5 mm) is the same in each photograph. Note the large grain-size range in the same specimen (10049) at left. The opaque mineral (black) is mainly ilmenite; the light-colored elongate crystals are plagioclase; and the gray, irregularly bounded grains are pyroxene.

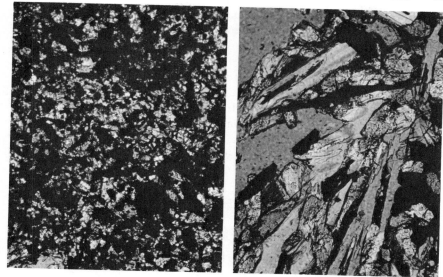

Fig. 4-3. Crystal-size ranges markedly in some of the type A igneous rocks. The width of the field is 0.3 mm in each photomicrograph, which are of different areas in the same thin section of specimen 10049. The opaque mineral (black) is ilmenite; the elongate gray crystals (visible only in the right photograph) are plagioclase; the gray irregular crystals with high relief (also distinct only in the right photograph) are pyroxene.

The type A and B rocks differ slightly in the abundance of ilmenite and plagioclase. The coarser-grained samples (type B) appear to be slightly higher in plagioclase and lower in ilmenite than the fine-grained samples (Mason and Melson, 1970). Also, the type B samples are lower in interstitial fine-grained material (mesostasis). The pyroxene contents, however, appear to be about the same. These mineralogical differences are reflected in the bulk chemical analyses by slightly higher alumina contents and slightly lower titania in the type B (coarse-grained) samples.

A wide variety of minor but genetically important minerals have been found in the Apollo 11 basalts, particularly in the coarser-grained samples. These include minute blebs of troilite and metallic iron (kamacite), which evidently crystallized from an iron-sulfur melt that was immiscible in the silicate liquid. A number of these minor minerals are associated with crystallization from the last several percent of liquid. These include the new iron-rich silicate pyroxferroite, which occurs on pyroxene margins and as separate grains, potassium feldspar, calcium phosphates (apatite and whitlockite), zirconium oxide (baddeleyite), and a rare earth-bearing calcium titanate (perovskite, variety dysanalyte). Among the minor early-crystallizing minerals is the new lunar mineral armalcolite (an iron titanium oxide), which occurs as cores in some ilmenite crystals. The properties of these minerals were discussed in Chapter 3 (lunar mineralogy).

Many of the Apollo 12 igneous rocks are porphyritic, that is, they contain large crystals of one or more minerals in a much finer-grained matrix. Large phenocrysts of pyroxene with pigeonitic pyroxene cores are a particularly noteworthy feature of some (Fig. 4-4). Others contain small phenocrysts of olivine in a matrix composed largely of skeletal quench crystals of olivine and feathery sheaths of pyroxene quench crystals (Fig. 4-5) without visible plagioclase.

Fig. 4-4. Some Apollo 12 igneous rocks contain large phenocrysts of pigeonitic pyroxene with hourglass structure. Crossed nicols. Width of field is 8 mm. NASA photomicrograph of sample 12021.

Fig. 4-5. Quench crystals of clinopyroxene (curved, feathery crystals in center of field) and olivine (long skeletal crystals) in an olivine-rich, plagioclase-poor igneous rock from Apollo 12 (12009). Width of field is 0.5 mm.

One of the most distinctive Apollo 12 samples is composed largely of plagioclase and alkali feldspar (rock 12013). The rock has had a complex origin and is extremely heterogeneous in thin section. In places, under low magnification, it appears to be a fragmental rock, somewhat like the microbreccias, but richer in large angular mineral fragments (Fig. 4-6). Under higher power, there is clear evidence of some recrystallization of the breccia, with development of a granulite texture (Fig. 4-6). This rock may be from an impact ejecta blanket which has undergone recrystallization while the ejecta were still hot, or it may be from a sheet of volcanic ejecta which underwent recrystallization after deposition. The rock may also have undergone contact metamorphism, either as an inclusion in magma or at the margin of an intrusion. The high feldspar content, along with certain chemical features, led to the classification of the rock as a feldspathic differentiate by the LSPET(1970).

CHEMICAL COMPOSITION

The Apollo 11 basalts have compositions quite unlike common terrestrial basalts or basaltic-type meteorites. The unusually high titania content first attracted attention to this; soon, numerous other chemical anomalies were

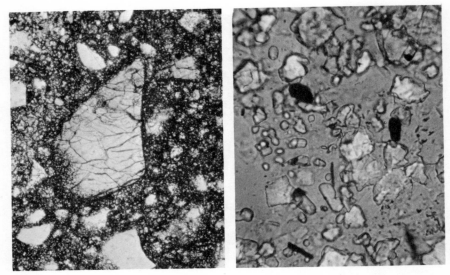

Fig. 4-6. Recrystallized feldspathic breccia (12013). This unusual sample has a well-developed fragmental texture in places (left, width of field is 0.5 mm) which under higher magnification shows a granulite texture (right, width of field is 0.03 mm).

found. Most of these anomalies were found in the Apollo 12 samples as well. Immediately, the high content of titania in the Apollo 11 samples was puzzling; indeed, although much work has now been done, there is still no good explanation. This, and other chemical features, are more fully discussed in Chapter 7.

The coarse and fine-grained Apollo 11 basalts have similar major and minor element compositions, but they have distinctly different compositions for some trace elements. In Table 4-1 we give selected analyses for the type A and B rocks. There are a number of unresolved differences in the reported analyses; for example, water has been reported present in very small amounts by some authors (e.g., Agrell et al.) but evidently was not detected by most of the other analysts. Indeed, measurements by other means have not indicated the presence of significant water contents. Also, some authors have reported the presence of ferric iron, whereas most indicate that it is totally absent.

As expected from the range in mineralogy, the Apollo 12 igneous rocks range considerably in chemical composition. Table 4-2 includes a number of preliminary analyses of selected varieties (LSPET, 1970). Also included is the highly feldspathic sample (12013), which is quite unlike all

Table 4-1. Selected Analyses by Classical Wet Methods of the Four Types of Apollo 11 Samples

	Igneous rocks			Microbreccia	Fines
	1 Type A 10022	2 Type B 10044	3 Av. A + B	4 Type C 10060	5 Type D 10084
SiO_2	40.53	42.46	40.38	41.96	42.16
Al_2O_3	8.52	10.21	9.43	11.85	13.60
FeO	19.76	17.60	19.32	16.51	15.34
MgO	7.68	5.96	7.20	7.63	7.76
CaO	10.42	12.25	11.05	11.38	11.94
Na_2O	0.54	0.48	0.46	0.49	0.47
K_2O	0.27	0.11	0.17	0.20	0.16
MnO	0.24	0.28	0.26	0.23	0.20
TiO_2	11.74	9.18	10.90	9.02	7.75
P_2O_5	0.14	0.04	0.12	0.07	0.05
Cr_2O_3	0.35	0.21	0.33	0.31	0.30
S	0.24	0.18	0.19	0.15	0.12
Fe	n.d.	0.60	0.20	0.60	0.60
	100.43	99.56	100.01	100.40	100.45

[1] L. C. Peck, analyst (1970, p. 584).

[2] J. H. Scoon, analyst (Agrell et al., 1970, p. 584). Also reported, but not found by other analysts, are $H_2O+ = 0.10$ and $H_2O- = 0.01$.

[3] Average of 11 selected type A and B crystalline rocks analyses by wet methods.

[4] Ibid. Also reported, but not found by other analysts, are $H_2O+ = 0.11$, and $H_2O- = 0.04$.

[5] Ibid. Also reported, but not found by other analysts, is $H_2O+ = 0.05$.

other Apollo samples in major and minor elements, as well as in trace elements. Comparisons between these and the Apollo 11 samples include (LSPET, 1970): (1) titanium contents are lower but are mainly still higher than those of common terrestrial rocks and meteorites, the range in TiO_2 being 1.2–5.1%, compared to 7–12% in the Apollo 11 samples; (2) potassium, zirconium, rubidium, lithium, yttrium, and barium are lower in the Apollo 12 rocks; (3) iron, magnesium, nickel, cobalt, vanadium, and scandium are higher in the igneous rocks from Apollo 12; and (4) significant variations occur between elements concentrated in ferromagnesium minerals, as the amount of these minerals ranges much more than in the Apollo 11 samples.

Table 4-2. Composition of Selected Apollo 12 Igneous Rocks and Their Average Composition (LSPET, 1970), and the Average Composition of Apollo 11 Igneous Rocks

	Apollo 12 rocks				Average	Apollo 11 rocks	Average
	1	2	3	4	5	6	7
SiO_2	35	41	40	49	40	61	40.6
Al_2O_3	11	11	12	12	11.2	12	9.5
FeO	23	20	22	17	21.3	10	19.4
MgO	17.5	12.5	8	6.5	11.7	6.0	7.2
CaO	9.3	10	12	11	10.7	6.3	11.1
Na_2O	0.53	0.51	0.42	0.60	0.45	0.69	0.46
K_2O	0.055	0.063	0.084	0.057	0.065	2.0	0.17
MnO	0.17	0.19	0.32	0.26	0.26	0.12	0.26
TiO	3.1	3.3	4.9	3.2	3.7	1.2	11.0
Cr_2O_3	0.57	0.76	0.44	0.32	0.55	0.15	0.33

1–4 Apollo 12 crystalline rocks (LSPET, 1970). The samples are, from left to right, numbers 12, 9, 64, and 38.

5 Average of 9 Apollo 12 crystalline rocks (LSPET, 1970).

6 Feldspathic (plagioclase and sanidine-rich) rock. Apollo 12 (12013); includes 0.30 ZrO_2 (LSPET, 1970).

7 Average of 11 crystalline rocks (types A and B) from Apollo 11.

CLASSIFICATION

Igneous rock classifications are based on a number of properties. These include mineralogical composition, texture (particularly grain size), chemical composition, and field occurrence. The particular name assigned to a rock on the basis of one property may differ from that based on another property. It is thus not surprising that a number of rock names have been applied to identical lunar samples.

The most widely used and practical classification of crystalline rocks is based on their mineralogical composition. The determination of mineralogical composition is readily made by estimates or more precise counts of the relative amounts of major minerals. The abundance of plagioclase and augite in the Apollo 12 samples places them in the basalt family. However, the abundance of ilmenite, and the combined abundance of ilmenite and pyroxene, has led them to be called ilmenite melabasalts following the mineralogical classification of Johannsen (1937); the prefix "mela" is used to point out this abundance of ilmenite and augite compared to normal basalts.

Depending on average grain size, the Apollo 11 basaltic rocks have been termed basalts (less than 0.5 mm), dolerites (greater than 0.5 mm but less than 2 mm), and gabbros (greater than 2 mm). The coarse-grained samples (type B) have also been termed microgabbros, which appear to correspond to the upper grain-size range of dolerites and the lower range of gabbros. Some of the Apollo 12 samples are much coarser and fall well within the gabbroic grain-size range.

Many of the Apollo 12 samples may be properly termed lunar basalts. Others have mineralogical compositions unlike basalts but analogous to other terrestrial rock types. These include pyroxene-rich peridotites (50% pyroxene, 40% olivine, and 10% plagioclase) and troctolites (15% pyroxene, 40% olivine, and 45% plagioclase) (LSPET, 1970).

One Apollo 12 sample is quite different from all the rest (analysis 6, Table 4-2). It consists largely of plagioclase, sanidine, and probably silica-group minerals. The silica content is the highest yet recorded for Apollo samples. This mineral has numerous other distinctive chemical properties; it has been termed a feldspathic differentiate of more mafic samples. Detailed investigations of its mineralogy are not at present completed, but we expect that it will add new members to this group and thus extend the list of known lunar minerals.

There are a number of classifications that are based solely on chemical composition. Chemical classifications are particularly useful in recognizing chemically identical rocks that have cooled at different rates and thus may have different mineralogical compositions and textures, or may be totally glass. One of the most widespread chemical classifications of terrestrial basaltic rocks depends on what is termed *normative mineralogy*. This classification involves the calculation of the amounts of a number of hypothetical minerals according to a definite sequence of steps. For basaltic rocks the amounts of these calculated minerals do not differ markedly from those that are in a totally crystallized sample. Based on normative mineralogy, basaltic rocks are often divided into oversaturated with regard to silica (quartz is a normative mineral), saturated (neither quartz nor olivine are normative minerals), and undersaturated (olivine is a normative mineral). The Apollo 11 samples are very close to saturation, that is, they contain only small amounts of either normative quartz or normative olivine. Most of the Apollo 12 samples are undersaturated, that is, they contain considerable normative olivine. An exception is the peculiar silica-rich, sanidine-plagioclase rock (12013), which contains about 20% normative quartz.

Another chemical classification of igneous rocks is based on the silica content. According to this scheme, most of the Apollo 11 and 12 igneous rocks would be classified as ultrabasic rocks, that is, they have silica contents of less than 45%. A few are slightly higher than this but have less than 55% silica; they are basic rocks. The highest silica rock yet recorded contains 61% silica and is thus an intermediate rock. Although no individual rocks have been found with higher silica contents than this, patches of rare residual glass in the type B basalts, as will be discussed below, reach over 66% silica and would be classified as acidic.

MAGMA VISCOSITY AND DENSITY

Viscosity refers roughly to the "fluidity" of a liquid. Examples of liquids with low viscosity include water and glycerine. Higher viscosities are shown by tar, by other thick fluids, and most terrestrial magmas. The viscosity of magmas is an important property. It determines, for example, flow thickness and the velocity of lavas on slopes. Also, in thick masses of cooling magma, viscosity determines, along with magma density and the density of any crystals the magma contains, how rapidly the crystallizing minerals will sink or float.

The viscosity of the parent magmas for the Apollo 11 igneous rocks has been calculated from the chemical composition and measured experimentally. Viscosity is a function of temperature; normally, the higher the temperature, the lower the magma viscosity. The viscosities of the lunar magmas in Mare Tranquillitatis evidently were low. Indeed, analogies with better-known liquids indicate that, at 1250° C, which is probably near the eruption temperature, they were probably as fluid as glycerine at room temperature. The actual calculated viscosity at 1250° C is about 27 poises (Weill et al., 1970), and the viscosity of a synthetic melt like the lunar basalts gave a viscosity of 7.1 poises at 1395° C (Murase and McBirney, 1970). This latter temperature is considerably higher than the likely lava extrusion temperature and thus the viscosity is probably lower than the actual value. At this temperature the density of the liquid was found to be about 2.95 g/cm^3.

This very low viscosity indicates that the lava flows would be very thin on slopes. Once an upper crust has formed, the lava can flow very long distances on even very gentle slopes. Thus the flows would have characteristics that would readily lead to the filling of topographic lows and eventual creation of broad, flat-floored areas, such as is true of portions of the maria. This low viscosity would also tend to give relatively inconspicuous vent areas, that is, the lava would rapidly move away from the vent (unless the vent was in the central part of a depression) and thus would not tend to build a volcanic edifice around it. Furthermore, crystal setting and floating will be greatly enhanced in lava lakes and large intrusions by this low viscosity, as will be discussed in the following section.

In silicate magmas the ease with which ions move toward crystal nuclei is related to viscosity. Low viscosity favors the rapid formation of crystals during cooling. Indeed, for very low viscosities, it is difficult to cool magmas rapidly enough to form glass. Thus the low viscosity of the lunar lavas accounts for, and is reflected by, the nearly totally crystalline nature of even the very fine-grained type A rocks.

CRYSTALLIZATION SEQUENCE OF LUNAR MAGMAS

When magmas cool from a totally molten state, minerals normally crystallize in a definite sequence that is determined by magma composition and pressure. At very high pressures, such as in the deep lunar interior, the minerals that crystallize are normally quite different from those which crystallize near or on the lunar surface. This section deals with the crystallization sequences of likely lunar magmas.

There are a number of reasons why the determination of the crystallization sequence of lunar magmas is so important. It provides clues about how a single lunar magma might be split, or differentiated, into a number of rock types. The mineral that crystallizes first will normally either sink or float in the magma if cooling is slow and there is a sufficient difference in density between crystals and magma. As these crystals are removed they will accumulate to form a rock, termed a *cumulate,* which is much enriched in this early-crystallizing mineral. As crystals separate, the remaining liquid becomes depleted in those elements in which the crystals are enriched, generating a new liquid composition. Melting experiments at high pressures limit the choice of likely parental materials for the lunar basalts and provide reasonable mechanisms for differentiation of basaltic liquids deep within the moon.

A number of experiments have determined crystallization sequences for the Apollo 11 samples. The results of some of those experiments at atmospheric pressure are summarized in Fig. 4-7. Investigators took precautions to prevent oxidation or further reduction of the melts. The temperature of complete melting is between 1210 and 1150° C, and of complete solidification, between 1050 and 1090° C, for these selected experiments. In all experiments, ilmenite or the iron-titanium oxide armalcolite [$(Fe, Mg) Ti_2O_5$] appeared early. Olivine, when present, crystallized before the remaining phases; plagioclase is one of the last phases to appear.

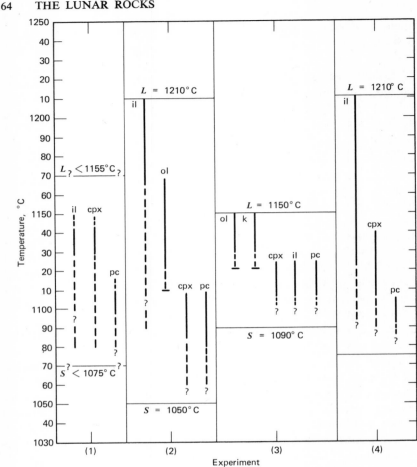

Fig. 4-7. The melting interval of the Apollo 11 "basalt" is relatively small. Temperatures of complete melting (liquidus, indicated by L) and of complete crystallization (solidus, indicated by S) of various basalt samples or their synthetic equivalents at 1 atm total pressure. Key: il = ilmenite, cpx = clinopyroxene, pc = plagioclase, ol = olivine, and k = karooite (armalcolite). (1) synthetic ferrobasalt (Anderson et al., 1970); (2) type B lunar basalt 10017 (O'Hara et al., 1970); (3) synthetic lunar model basalt (Ringwood and Essene, 1970); and (4) melting and crystallization of natural glass inclusions in various Apollo 11 basalts. (Roedder and Weiblen, 1970a.)

The differences in the experimental results may reflect slight differences in the composition of the starting materials.

Crystallization temperatures have also been estimated by measuring oxygen isotope fractionation between coexisting plagioclase and ilmenite in a number of Apollo 11 basalts. However, a number of assumptions were required to calculate these temperatures, mainly because experimental work on ilmenite-plagioclase partitioning has not been done. Epstein and Taylor (1970) calculated that ilmenite and plagioclase crystallized at temperatures in the range of 1338–1154° C for five crystalline rocks. These are higher than the values obtained in melting experiments (Fig. 4-7). Somewhat lower temperatures around 1120° C were estimated by Onuma et al. (1970).

These temperatures of isotopic equilibration are much higher than those measured by oxygen isotope fractionation between minerals in terrestrial gabbros. This has been attributed to lack of low temperature reequilibration in the coarse-grained lunar samples during cooling, perhaps because of the absence of significant H_2O during cooling of the lunar basalts. There is still another explanation: the very low viscosity of Apollo 11 basaltic magma leads to rapid and coarse crystallization, even when the magma was rapidly cooled. Thus most terrestrial basaltic magmas, which are more viscous, would have to cool much slower than a lunar magma to produce comparable grain size; hence there is more time for reequilibration.

The experimentally determined crystallization sequence agrees well with that indicated by microscopic features. Ilmenite appears to be the first mineral to crystallize. It forms large crystals in the type B samples (Fig. 3-3) and has cores that rarely enclose other minerals. Olivine, in those few samples in which it is present, forms relatively inclusion-free crystals that are rimmed and partially replaced by augite. Augite appears to crystallize somewhat later than olivine but in part precedes the crystallization of plagioclase. The well-developed ophitic texture (Fig. 2-5) in the type B basalts reflects the simultaneous crystallization of plagioclase and augite.

The observed early crystallization of the dense minerals ilmenite and olivine under low pressure suggests a number of reasonable low-pressure differentiation trends. In large magma chambers, or in deep lava lakes, the settling of ilmenite, and olivine if present, will occur and produce dense basalt layers. Although settling rates would be slowed by the low lunar gravity, which is but one-sixth that of the Earth, the low viscosity of the parent magma would more than compensate for the lower gravity. For example, an ilmenite crystal about 1 cm across would settle at about 2 cm/sec. On Earth, a similar crystal in basaltic magma, compositionally like the Columbia River basalts, would settle at only 0.5 cm/sec because of the higher viscosity. Allowing even short cooling times, on the order of days, major separation of ilmenite would occur. Indeed, it is difficult to imagine how any slowly cooled lunar basaltic magma could retain much of its ilmenite. With ilmenite removal, magma viscosity would increase because of increasing silica content and lower temperatures, and crystal sinking (or floating) would be slowed.

66   THE LUNAR ROCKS

Like terrestrial basalts the crystallization sequence of the Apollo 11 basalt changes drastically with pressure. The crystallization sequences may be divided into three different pressure regimes (Fig. 4-8; after Ringwood and Essene, 1970). Ilmenite, armalcolite (karooite), and olivine are typically first phases to crystallize at pressures up to 3 kb, which corresponds to a depth of about 60 km within the Moon. At greater depths, in the range of 200 to 400 km, only clinopyroxene (subcalcic augite, $Fs_{21}En_{64}Wo_{15}$, with $Al_2O_3 = 4\%$, $CaO = 7\%$) and sometimes orthopyroxene ($Fs_{20}En_{74}Wo_6$) appear on the liquidus. Garnet and clinopyroxene become the liquidus phases in the highest pressure regime, that is, at pressures greater than 25 kb, corresponding to depths greater than 550 km within the Moon.

The Apollo 12 igneous rocks do not have a simple crystallization sequence based on their textures. Phenocrysts of pigeonite occur in some, of olivine in others. These presumably are the low-pressure liquidus phases of their parent magmas. Like the Apollo 11 basalts, most of these rocks are low in silica; thus their parent magmas probably also had low viscosities. The lower content of ilmenite and the greater abundance of

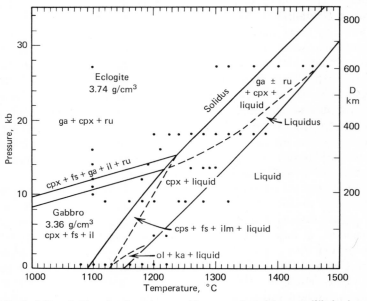

Fig. 4-8. Stability fields of mineral assemblages and melting equilibria in average Apollo 11 synthetic basalt at high pressures and temperatures. Each dot represents a separate experiment. cpx = clinopyroxene, fs = feldspar, ga = garnet, il = ilmenite, ru = rutile, ka = karooite, and ol = olivine. (Ringwood and Essene, 1970).

olivine are also notable differences. One might speculate that the low ilmenite content of the Apollo 12 samples is a result of near-surface partial crystallization, during which much ilmenite was removed by settling to the bottom of magma chambers.

At present we have no information on the experimental behavior of the diverse Apollo 12 igneous rocks at high temperatures and pressures. One might expect, though, that the more mafic, olivine-rich varieties melt totally at considerably higher temperatures than the Apollo 11 samples. However, the abundance of large crystals of dense phases, such as olivine, raises the possibility that the samples do not represent liquid compositions, that is, they may have become enriched in the phenocryst minerals derived from elsewhere. For example, the olivine-rich samples may be from the bottom of magma chambers, where olivine crystals had accumulated from a large volume of overlying crystallizing magma. Such samples are common in terrestrial basaltic volcanic sequences, and the problems they produce when viewed as representative of liquid compositions, such as unreasonably high melting temperatures, have long plagued petrologists.

RESIDUAL GLASSES AND LIQUID IMMISCIBILITY

The residual glasses in the Apollo 11 basalts contain two glass phases: (1) a high-silica, low-potash glass, akin to granitic or rhyolitic compositions (analyses 1–6, Table 4-3), and (2) a low-silica, high-FeO glass (analysis 7, Table 4-3). In some samples, these glasses form globules of one within the other. This relationship, of one glass within another, indicates that the glasses were derived from immiscible liquids, that is, from liquids which, like oil and water, do not mix. The discovery of this silicate liquid immiscibility and its implications on the origins of lunar magmas is due to the work of Roedder and Weiblen (1970a, 1970b).

The splitting of these two liquids occurred when the rocks were almost solidified, when only about 2–10% liquid remained. Thus the immiscibility affected only a small amount of material. Nevertheless, it shows that the Apollo 11 basaltic parent magmas could produce small amounts of both a granitic liquid and a very iron-rich liquid if allowed to crystallize slowly. If the cooling and crystallization took place at depth, this granitic residual liquid might be separated from the parent magma and ascend to form either granitic intrusions or rhyolitic lavas on the lunar surface. One would also expect to find the complementary low-silica, iron-rich liquid, which indeed has a most peculiar composition for known lunar and terrestrial igneous rocks. The presence of residual liquids of rhyolitic com-

Table 4-3. Analyses of Some Residual Glasses in the Apollo 11 Basalts

	1	2	3	4	5	6	7
SiO_2	78.92	73	62	78.1	71	76.1	47.8
TiO_2	...	1.2	0.35	1.11	1.1	0.5	3.7
Al_2O_3	11.88	13	23	10.6	11.5	11.7	3.2
Cr_2O_3	...	...	...	0.05	...	...	...
FeO	2.88	2.4	3	0.96	6.1	2.5	31.4
MnO	...	0.2	...	0.04	0.1	...	...
MgO	0.08	0.2	0.1	0.07	0.2	0.3	2.3
CaO	2.10	2.1	9	5.1	2.4	1.9	11.2
Na_2O	0.56	0.5	1.7	1.24	0.7	0.4	0.1
K_2O	6.86	5.5	2.3	3.4	6.6	6.6	0.3
P_2O_5	...	0.3	0.09	0.19	...	...	...
ZrO_2	...	...	...	0.04	...	...	...
	103.28	98.4	101.54	100.90	99.7	100.0	100.0

[1] Agrell et al. (1970), interstitial glass. Table 2, analysis 13.

[2] Anderson et al. (1970). Table 1, analysis D. Devitrified glass fragments of ferrobasalt in breccia 10061.

[3] Ibid., Table 1, analysis C. Devitrified glass in ferrobasalt 10022.

[4] Keil et al. (1970). Igneous glass in minute patches, less than 30 μm in diameter, with SiO_2 range from 72–82%. Typical analysis is given.

[5] Kushiro et al. (1970). One of ten analyses of mesostasis in sample 10024 (coarse-grained lunar basalt).

[6] Roedder and Weiblen (1970a). Composition of high-silica liquid coexisting with low silica liquid. Analyses range from rock to rock; weighted average of 37 analyses is given.

[7] Ibid. Low-silica glass. Ranges in composition from sample to sample. Weighted average of 37 analyses is given.

positions and of silica contents intermediate between rhyolites and basalts. Andesitic glasses were noted by a number of other workers (analysis 3, Table 4-3). Commonly, these, too, were taken to indicate that the Apollo 11 basaltic magmas had the potential under ideal conditions to generate rhyolitic and andesitic liquids. As will be discussed in Chapter 8, this potential was viewed as one way of generating postulated feldspathic lunar highlands and of providing the parent material for tektites.

Associated with the residual glass are concentrations of minor minerals. Like the residual granitic glass, these attest to the very strong chemical fractionation in the late-stage liquids. These minor phases include silica, as cristobalite (Fig. 4-9) and more rarely tridymite; potassium feldspar; calcium phosphates; zirconium silicate; and zirconium oxide. Barium is highly fractionated into the potassium feldspar. Zirconium appears to be about two-thirds in the oxide and silicate, and about one-third in ilmenite.

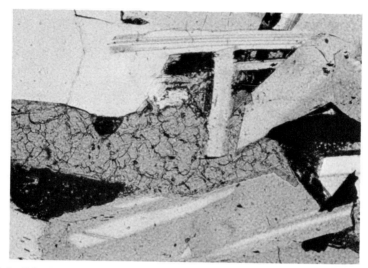

Fig. 4-9. Cristobalite is a late phase to crystallize from the Apollo 11 basaltic magma. Cristobalite (mineral in center left with abundant curving fractures) is in the interstices between plagioclase crystals. The dark material is very fine-grained mixture of various phases, including two immiscible glasses (not visible), the new pyroxenoid pyroxferroite (also not separately visible), and other minor phases. The width of the field is 0.5 mm; coarse-grained (type B) rock 10047.

The rare-earth elements appear to be greatly enriched in the phosphate minerals and in the high-silica glass.

LUNAR ROCK MAGNETISM

Like most terrestrial rocks, the lunar samples possess distinctive and informative magnetic properties that are readily measurable and reflect their past history. These properties may be divided into two closely related categories: The first is determined predominantly by the mineralogy and grain size of the sample; the second is magnetizations acquired during the history of the rock sample. The intensity of the past or remnant magnetization is determined by rock mineralogy and grain size as well as past history. Nagata (1961) has presented a detailed review of the principles of rock magnetism.

The presence of magnetic properties in a rock may be shown, for example, by suspending a small fragment of the rock in a very strong magnetic field such as that generated by an electromagnet. Substances like metallic iron are strongly affected by such a magnetic field and are termed *ferromagnetic*. The small amounts of metallic iron in the Apollo 11 samples account for many of their magnetic properties and are the main carriers of the remanent magnetization. Some substances are only weakly attracted in magnetic fields and are termed *paramagnetic*. The major paramagnetic mineral in the lunar samples is pyroxene, which owes its paramagnetism to the presence of a small number of free ferrous ions. Some minerals have properties like paramagnetic substances, except that they become strongly magnetic only at a given temperature; the magnetization falls off rapidly both above and below this temperature. Such minerals are called *antiferromagnetic;* in the lunar samples they include ilmenite and the ferrosilite ($FeSiO_3$) molecule in the pyroxenes. Both these substances become strongly paramagnetic only at very low temperatures and thus in practice do not play important roles in the magnetic properties of the lunar samples.

The behavior of some lunar crystalline rocks when placed in a magnetic field and magnetized at very low temperatures has been shown to be a predictable sum of the contributions of the free ferrous ions in pyroxene, the ferrosilite molecule, ilmenite, and metallic iron (Nagata et al., 1970). A rock with these substances in the above order and in the ratio of 4.3–7–20–0.08 wt %, can closely simulate the observed magnetic properties of one of the lunar type B crystalline igneous rocks (Fig. 4-10). These ratios are similar to those expected from optical studies.

The abundance of the metallic iron particles and their nickel content also have been inferred from the thermomagnetic experiments. Strangway et al. (1970) suggest that these particles include low-nickel iron, iron with 5–7% nickel, and iron with greater than 33% nickel. This range in nickel contents is confirmed by direct electron-probe analyses. The magnetic studies indicate the presence of about 0.3–0.5% total metallic particles, which agrees well with the amount determined by chemical analyses: 0.20–0.60% in samples of the fines, a breccia, and two lunar basalts (Agrell et al., 1970).

Remanent magnetization is acquired, among other ways, when a magnetic mineral cools through its Curie temperature in the presence of a magnetic field. On cooling, an igneous rock that contains minerals such as metallic iron acquires a magnetization of its own which is parallel to the magnetizing field and may remain substantially unaltered for billions of years after it is acquired. On Earth, studies of remanent magnetization of rocks have provided key support for the now widely accepted theories of sea-floor spreading and continental drift.

Scientists expected such studies on the lunar samples to reveal evidence concerning the Moon's past magnetic field. Satellite measurements had shown that the Moon does not now possess a detectable magnetic field.

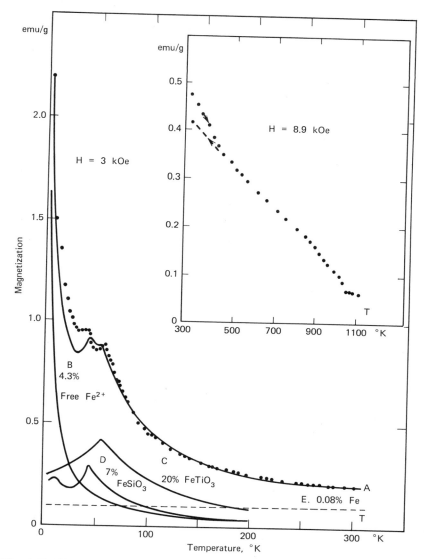

Fig. 4-10. The magnetization of a lunar sample is the sum of the magnetization of the individual constituents. Magnetization is shown as a function of temperature in a field 3000 Oe for a coarsely crystalline lunar basalt (sample 10024). Solid circles are the measured values; solid lines are simulated values and individual values for the dependence of the net magnetization (curve A) of free Fe^{+2} in pyroxene (curve B), ilmenite ($FeTiO_3$, curve C), $FeSiO_3$ (ferrosilite molecule in pyroxene, curve D), and of metallic iron (curve E). The insert shows a measured magnetization at higher temperatures in a magnetizing field of 8900 Oe. (Modified from Nagata et al., 1970.)

Did it have one in the past? Hopefully the absence or presence of remanent magnetization in the lunar sample would contribute to this question. The Earth's magnetic field is due to the presence of an electrically conducting fluid core (Malkus, 1968). If the lunar samples possess a remanent magnetization, this would imply that at some time in the past the Moon also had a fluid conducting core, although, from the known overall low lunar density, it would have had to be considerably smaller than the earth's core and have been made up of a smaller part of the lunar mass.

Measurements of the remanent magnetization of the Apollo 11 samples were made by a number of investigators (Table 4-4). These results, which differ, reflect intrinsic differences in the samples rather than analytical error. Each investigator examined a different group of samples, and, based on their properties, drew conclusions about possible lunar history. Since the results of their measurements differed, their conclusions also differed. Some found that certain samples possessed a strong and relatively stable remanent magnetization (Fig. 4-11) and concluded that a magnetic field had existed at the time the particular sample was formed. However, this remanent magnetization was not taken to prove the existence of a past regional lunar magnetic field. It was pointed out that the magnetic field may have originated in other ways such as (1) spurious very high intensity magnetic fields in the spacecraft (which were deemed unlikely), (2) continuous thermal cycling in a weak magnetic field generated by the solar wind, (3) instantaneous currents and magnetic field generated by large impact events, and (4) a result of cooling in the realm of the Earth's magnetic field when the Earth and Moon were possibly much closer together than they are at present.

The divergence in the results of the measurements of the remanent magnetizations raise doubts about the existence of a lunar magnetic field even 3.7 billion years ago. One might expect, for example, a more uniform remanent magnetization in the lunar "basalts" if they had all crystallized in a magnetic field, as they have similar metallic iron contents. Two fine-grained lunar "basalts" (type A) were found to have a reasonably stable remanent magnetization (Fig. 4-11) by Helsley (1970). Strangway et al. (1970) attribute the lack of remanent magnetization in some of the lunar breccias they measured to either demagnetization by shock or to the absence of a magnetic field when they crystallized. If shock were responsible for the demagnetization of some of the lunar basalts, one might find a correlation between the textural evidence of shock and the intensity of remanent magnetization. So far, such a correlation evidently has not been sought.

Nagata et al. (1970) cast much doubt on the idea that their crystalline

Table 4-4. Selected Results of Natural Remanent Magnetization (NRM) Measurements of Apollo 11 Samples. Not All the Apollo 11 Samples Possess a Significant NRM

Sample type and number	Investigator	NRM (emu/gm)*	Comments on results
Lunar basalt (type B) 10024, 22	Nagata et al. (1970)	7.5×10^{-6}	Magnetization direction and intensity different in each of five slices; therefore, NRM is not a thermal remanent magnetization. NRM is due to ferromagnetism of fine metallic iron.
Lunar basalt (type B) 10017, 64	Runcorn et al. (1970)	5.6×10^{-6}	Relatively stable NRM. This is perhaps a thermal remanent magnetization reflecting past presence of a lunar magnetic field or may reflect thermal cycling in the presence of a magnetic component of the solar wind.
Lunar basalts (type A) from dust sample 10085, 6; and sample 10069	Helsley (1970)	2.46×10^{-5} 1.92×10^{-5}	Relatively stable NRM. This NRM may be interpreted as a thermal remanent magnetization acquired by cooling in a field in excess of 1500 gammas (Earth's magnetic field ranges from 30,000–70,000 gammas).
Microbreccia. 10059, 24	Doell et al. (1970)	5.3×10^{-5}	Stable NRM; possibly acquired by impact process on the lunar surface.
Microbreccia. 10046, 46	Runcorn et al. (1970)	6.2×10^{-5}	Unstable; "hard" remanence is very low, less than 1.0×10^{-6} emu/g. "Hard" magnetic particles that it and the lunar fines contain are most likely randomly oriented single domains. Bulk susceptibility is 2.9×10^{-3} emu/g, susceptibility anisotropy less than 0.5%.
Microbreccia. 10023	Strangway et al. (1970)	2.8×10^{-3}	Not a stable magnetization but is typical of many rocks deemed adequate for paleomagnetic studies. This magnetization could only have been obtained in a magnetic field, but this field may have originated in a number of ways.
Microbreccia. 10048, 22	Larochelle and Schwarz (1970)	3.7×10^{-4}	Very low stability; not a good carrier for paleomagnetism.

* Electromagnetic units per gram

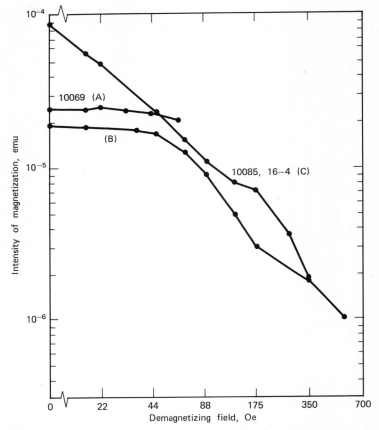

Fig. 4-11. Remanent magnetization can be erased by placement in strong alternating magnetic fields. Some lunar samples show a stable natural remanent magnetization (NRM) (samples A and B, fine-grained lunar basalts), that is, an NRM that can only be erased at high demagnetizing field strength. Sample C (a vesicular basalt fragment from fines 10085) has no stable NRM and shows only an induced remanent magnetization. (Modified from Helsley, 1970.)

rock possessed a thermal remanent magnetization by an ingenious experiment. Although they detected a bulk remanent magnetization, they found that its direction differed in each of five slices of the same sample; that is, they found that the remanent magnetization was not homogeneously distributed through the rock, as would be the case for stable thermal remanent magnetization. They attribute the remanent magnetization to the ferromagnetism of fine metallic iron. They also speculate that the remanence may be a result of a pressure (piezomagnetic) effect caused

THE IGNEOUS ROCKS 75

by shock waves generated by meteorite impact or a simple heterogeneous distribution of minute single-domain particles of metallic iron.

An experiment showed that the lunar fines possess metallic grains having a "hard" (highly stable) remanent magnetization (Fig. 4-12). Run-

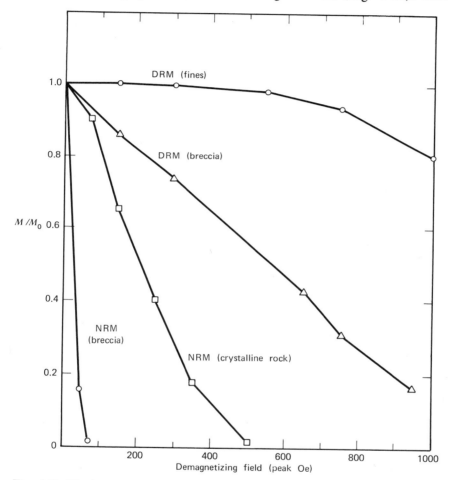

Fig. 4-12. The lunar dust contains minute metallic iron particles with a highly stable remanent magnetization. The relative stability of fines from the dust (circles) and a crushed breccia (triangles) were measured by their depositional remanent magnetization (DRM), acquired by allowing them to settle in acetone under the influence of a weak magnetic field. Also shown are the relatively low stabilities of the natural remanent magnetization of a crystalline rock (squares) and microbreccia (circles) sample. M/M_o is the ratio of the partially demagnetized intensity to the original intensity. (From Runcorn et al., 1970.)

corn et al. (1970) allowed a sample of the lunar "soil" and a crushed breccia sample to settle out in acetone in the presence of a weak magnetic field. This field tended to align the magnetic particles as they settled so as to create a so-called *depositional remanent magnetization* in the settled material. They then measured the stability of this depositional remanent magnetization and found that the fines, and to a much less extent the crushed breccia, contained a significant amount of particles with extremely high stability. They attribute this hard magnetization to the presence of single-domain iron grains that were randomly oriented in the undisturbed soil.

The remanent magnetization measurements raise many questions. The possibility that the Moon possessed a magnetic field and hence a fluid conducting core has neither been proven nor ruled out. Helsley (1970) estimates that there may have been a past field of perhaps a minimum of three-hundredths that of the Earth's present field. The role of impact processes in creating or erasing remanent magnetization in the lunar samples is not clear. On Earth, it has been found that shock both reduces and creates remanent magnetizations, depending on various factors (Hargraves and Perkins, 1969). Plans call for the eventual return of oriented samples from lunar bedrock exposures. Hopefully, these may help to resolve better the problem of a past magnetic field, as well as to give the direction in which any ancient field may have been oriented.

AGE

The ages of the lunar samples are inferred by radiometric means, that is, they are based on the decay of radioactive elements. The lead-uranium-thorium, rubidium-strontium, and potassium-argon methods have been used. Most of these methods give similar results. Thus the inferred ages are viewed as reliable indicators of the time of crystallization of the magmas from which the lunar crystalline rocks were derived. The ages of the microbreccias and regolith samples are more difficult to interpret, since they are a mixture of fragments of many different rocks that must be much older than the large igneous rock samples. This will be further discussed in the next chapter.

Some of the most important conclusions from the radiometric dating of the crystalline rocks are (1) their extreme ages, which cluster about 3.7 billion years, and (2) the fact that they are not all the same age. A number of investigators found readily measurable age differences between the samples. The oldest crystalline rock is a small fragment from the fines, which gave an age of 4.4 billion years. This 4.4-billion-year-old fragment differs from the larger samples and consists mainly of shock metamorphosed low-calcium pyroxene, isotropic "plagioclase" (maskelynite), potassium feldspar, ilmenite, troilite, and whitlockite. This peculiar fragment may have come from beyond the mare (Albee et al., 1970). Age differences between some of the larger samples are small, although distinctly measurable (Tatsumoto and Rosholt, 1970; Silver, 1970).

The Apollo 12 igneous rocks are considerably younger than the igneous rocks from Tranquillity Base, and are perhaps about 2.7 billion years old, based on the potassium-argon method (LSPET, 1970). This extremely important discovery indicates that lunar volcanism spanned a considerable part of early lunar history. This apparently younger age for the Apollo 12 samples awaits verification by the more reliable lead-uranium-thorium and rubidium-strontium methods.

ORIGIN OF LUNAR MAGMAS

Igneous rocks have now been found to be the most abundant rock type at two mare landing sites. What caused the melting that produced their parent magmas? Hypotheses for the origin of the melts may be classified into at least two groups: (1) melting caused by the impact of gigantic meteorites, and (2) melting in the lunar interior. Melting of the lunar interior may be attributed to at least three processes: (1) heat released from gravitational energy of the accumulating "planetesimals" during the accretion of the Moon (Baldwin, 1963), (2) heat from the decay of short-lived radioisotopes, such as aluminum 26, and (3) heat from the decay of the long-lived radioisotopes potassium 40 and those of the uranium and thorium series. The first two heat sources could have been important only during the early history of the Moon. The third source would be important throughout lunar history.

We believe it unlikely that the magmas which gave rise to the Apollo 11 and 12 rocks were generated by meteorite impacts. Such melts would normally crystallize over a very large temperature interval, whereas the lunar samples appear to crystallize over a narrow interval. Also, impact melts are commonly heterogeneous and contain abundant inclusions of only partially melted materials. These characters are found in lunar glasses in the regolith and soil but not in the igneous rocks.

The most likely hypothesis to us is that the lunar magmas at the Apollo sites were formed by melting in the lunar interior—a melting that was unrelated to meteorite impact. Anderson and Phinney (1967) calculated the rise in temperatures in the lunar interior due to heating by long-lived radioisotopes of an initially cool (273°K) Moon. They assume two abundances for these radioisotopes—one like that of chondritic meteorites and one like that of the Earth's mantle. They conclude that melting would have begun about 2.4 billion years ago for a chondritic Moon, and about 2.9 billion years ago for a Moon like the Earth's mantle. The preliminary ages of the lunar rocks at the Apollo 12 site (2.7 billion years; LSPET, 1970) indicate that melting did occur within this predicted interval. However, the age of the Apollo 11 basaltic rocks (3.7 billion years and older) indicate that melting occurred long before these predicted times. This may mean that (1) residual heat of accretion was higher than they assume, (2) short-lived radioisotopes played a more important role than assumed in their calculations, (3) the Moon is much older than the assumed 4.5 billion years, or (4) the content of long-lived radioisotopes is greater than either chondritic or terrestrial models.

Ideas on the thermal history of the lunar interior clearly have been greatly changed by the Apollo samples. The extensive revision of these ideas is under way. Furthermore, the possible role of large heat additions by a much hotter early Sun, which heated the accreting Moon, is now being seriously considered. However, the problem of having very old igneous rocks in the solar system is not unique to the Moon. Such old igneous rocks have not been preserved on Earth, but such ages are characteristic of the achondritic meteorites, some of which were clearly derived from magmas and are 4.5 billion years old.

Ringwood and Essene (1970) contend from their high-pressure, high-temperature experiments that the most likely parental materials underwent partial melting at a depth between 200 and 400 km, and that the parent rock was mainly subcalcic augite and orthopyroxene. Quite a different parent material has been proposed from the rare-earth element contents. Haskin et al. (1970) postulate that the likely parent rock contained at least 25% plagioclase, and no more than a few percent clinopyroxene. Such a parent rock could exist only at very shallow depths within the Moon because at high pressures plagioclase becomes unstable and reacts to form new minerals (Fig. 4-8).

Some process led to the slight differences in titanium and aluminum contents and major differences in the trace element contents (Chapter 6) between the type A and type B basaltic rocks. The slight enrichment in plagioclase in the coarser-grained samples is not readily accounted for by simple crystal settling of early-forming ilmenite (Mason and Melson, 1970). Rather, it appears accountable for by settling of ilmenite as well as some pyroxene. The subtraction of last liquids from slowly crystallizing type B samples may account for many of the trace element differences because the pertinent trace elements are concentrated in the mesostasis. Some combination of (1) settling of some of the early-crystallizing ilmenite and pyroxene, and (2) subtraction of the residual liquids, perhaps by "filter-pressing," is one scheme by which a magma of type A composition could produce type B rocks.

CHAPTER 5

LUNAR PETROLOGY: THE FINES AND MICROBRECCIAS

INTRODUCTION

The origin of lunar craters has been much debated. Most scientists believe that they were produced mainly by the impact of meteorites, although some believe that they are mainly volcanic. Impact processes and volcanism as we know them on Earth produce distinctly different rocks, and thus the Apollo samples provide keys to how lunar craters were produced. The lunar basaltic rocks attest to the major role volcanic eruptions had in flooring Mare Tranquillitatis. Although specific large craters would not be examined in the Apollo 11 and 12 missions, nevertheless it was anticipated that characteristic impact-generated debris might be found mantling the maria. Some of the lunar samples that show unmistakable evidence of impact processes are the lunar fines and the microbreccias.

Many of the lunar materials show features attributable to shock. These include (1) the lunar regolith, which is composed mainly of fragmental "throw-out" from impact craters; (2) the microbreccias, which may have been lithified by shock or by welding in base-surge deposits created by impacts; (3) glass spherules (Fig. 5-1), broken glass spherules (Fig. 5-2), and angular glass fragments, all of widely ranging compositions; (4) glass splashed on rock surfaces; (5) nickel-iron spherules, presumably fragments of impacting meteorites in the lunar soil and in some of the glasses; and (6) shock deformation features in minerals, including generation of planar structures, and decrease in birefringence.

LUNAR REGOLITH

At the Apollo 11 site the lunar surface is underlaid by a weakly coherent fragmental material, termed the *lunar regolith,* estimated to range be-

THE FINES AND MICROBRECCIAS 81

Fig. 5-1. The lunar regolith or "soil" contains a wide variety of glass particles. The average size of these particles is about 0.2 mm. (Photograph by E. C .T. Chao, U.S. Geological Survey.)

tween 3–6 m thick and average about 4 m thick. Underneath this loose material, there appears to be a more coherent surface, which may well be the surface of a 3.7-billion-year-old lava flow. This upper layer of loose material differs fundamentally from terrestrial soil. For this reason, some have objected strongly to calling the lunar regolith the lunar "soil." Regolith is a less specific term, referring only to a loose mantle of surficial material of unspecified origin.

Terrestrial soils are, like the lunar regolith, formed by the breakdown of coherent rock masses. However, the main agents causing the breakdown of rocks on Earth—such as the chemical and mechanical action of water—are totally lacking in the lunar environment. On the Moon, impacting bodies appear to play the major part in the formation of the lunar regolith. Such a process does not play a significant role in generating terrestrial regolith.

The thickness of the lunar regolith at Tranquillity Base, estimated at about 4 m, and the estimated age of the basaltic basement, about 4 billion years, indicate that the average rate of regolith formation is on the order

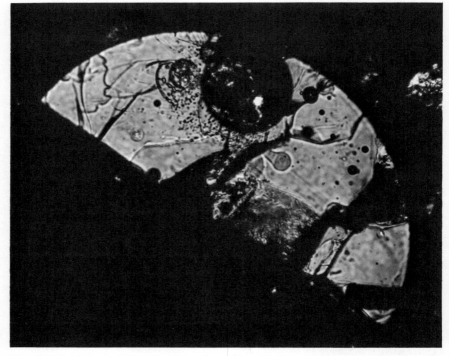

Fig. 5-2. Broken, dark brown glass spherule in microbreccia from the lunar regolith. The hemisphere is about 0.1 mm in diameter.

of 1 millimeter per million years. This is a remarkably slow average rate of accumulation, compared to even the lowest terrestrial sedimentation rates. For example, in the deep sea, very remote from continent sediment sources, the rate appears to average between 2000–5000 millimeters per million years. Of course, the lunar regolith at a given place may form almost instantaneously from debris ejected from a nearby impact crater. However, such blanketing occurs extremely rarely.

Aldrin, Armstrong, and Collins (NASA, 1969) divided the regolitn into four zones based on their extravehicular activity:

1. A very thin (less than ⅛ in.) light gray to tannish gray dusty zone.
2. A thin (about ¼ in.) dark gray caked zone that will crack 5–6 in. away from where force is applied.
3. A 2- to 6-in. zone of soft, dark gray to cocoa gray, slightly cohesive sandy to silty material that will hold slopes of probably 70°.
4. A zone that is gradational with zone 3, except for a marked increase in firmness and resistance to penetration.

Zone 3 was found to be thickest on the rims of small craters, whereas very little thickness variations were noted in the two upper zones.

The lunar regolith appears to have been penetrated totally in one of the larger craters visited by Armstrong. The regolith is excavated to a depth of around 4 m, and the crater contains a central rise of coarse broken rock, presumably from immediately underlying bedrock.

The two cores obtained from the regolith showed no obvious grain-size stratification. The cores penetrated 10 and 13.5 cm into the regolith before they were stopped by a more coherent layer. The material in the core tubes was quite porous, having bulk densities of 1.66 and 1.54 g/cm^3. There were some slight differences between the materials within the deeper core; near the top, they were slightly lighter colored, although in grain size and texture they did not otherwise differ from the deeper portions of the core.

The samples of the lunar regolith, termed the *lunar fines*, consist mainly of particles of plagioclase, clinopyroxene, ilmenite, olivine, and a wide variety of glasses and rocks. These particles are quite small, most corresponding in size to terrestrial coarse silt and fine sand (Fig. 5-3).

The lunar regolith is subject to a slow but definite turnover. This movement and stirring action comes about largely from cratering, both on a small and large scale. Shoemaker et al. (1970) have estimated from the distribution and sizes of craters at Tranquillity Base that the regolith is turned over to depth of about 4 m—the main thickness of the regolith

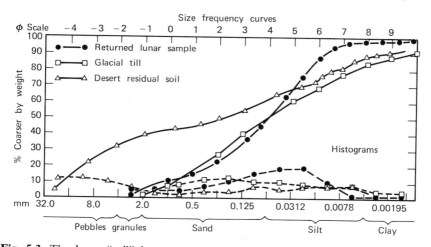

Fig. 5-3. The lunar "soil" is composed mainly of fragments that are like silt and fine sand, although larger and smaller fragments occur. Cumulative size distributions and histograms for the lunar sample and, for comparison, glacial till and desert residual soil. (Duke et al., 1970.)

84   THE LUNAR ROCKS

there—perhaps every 3.7 billion years, the estimated age of the underlying basaltic bedrock. Funkhouser et al. (1970) found from cosmic-ray exposure ages that individual rock samples were exposed at the surface from 20–400 million years. They noted that the ages fall into two rather broad groups, centered at about 100 million and 350 millions years, possibly reflecting the excavation and scattering of the rocks at Tranquillity Base by relatively few, distinct impact events.

The lunar fines contain a remarkable abundance and variety of glass particles. The preliminary examination team (LSPET, 1969) recognized the following types of glasses: (1) vesicular, and globular dark gray fragments; (2) angular pale or colorless and, more rarely, brown, yellow, or orange fragments with an index of refraction ranging from 1.5 to 1.6; and (3) spheroidal, ellipsoidal, dumbbell-shaped and tear-drop-shaped bodies, most of which are smaller than 0.2 m and range from red to brown and from green to yellow (Fig. 5-1). The index of refraction ranges from less than 1.6 to more than 1.8 and is generally higher for the more intensely colored glass. Unlike normal quenched magma droplets and glasses from terrestrial volcanic sources, many single glass particles in the core and bulk samples are inhomogeneous. The properties and origins of lunar glasses are discussed in more detail later in this chapter.

The samples of the regolith are similar in composition to the microbreccias (Table 4-1), but differ from the Tranquillity Base basaltic rocks, among other ways, in higher nickel and aluminum, and lower titanium content (Tables 7-1 and 4-1). These differences appear to reflect enrichment of the fines and microbreccias in plagioclase-rich rock fragments and meteoritic material. Fragments derived by the breaking and melting of bedrock, which is like the Apollo 11 basalts, are by far the most abundant constituents in the fines and in the microbreccias.

The Regolith as a Radiation Monitor

The lunar surface is subjected to continual bombardment with a variety of nuclear particles, from which the Earth is shielded by its deep atmosphere and its magnetic field. The principal bombarding particles are derived from the Sun (in the form of the solar wind and solar flares) and from cosmic rays. Particles in the solar wind and solar flares have comparatively low energies and hence low penetrating power, less than a millimeter in mineral grains. Cosmic-ray particles have high energies and corresponding high penetrating power, up to many centimeters in solids.

The solar wind and solar flare effects are largely responsible for the high gas content of the lunar breccias and fines in comparison to the crystalline rocks (Table 7-4). The effects of cosmic-ray bombardment are evidenced by the presence of their spallation products in the lunar rocks; cosmic-ray collision with atomic nuclei in rocks causes disintegration, with the production of a variety of unusual isotopes, some of them radioactive. Perhaps the most informative method for the study of these irradiation effects is the nuclear-track technique (Fleischer et al., 1969). In this procedure the narrow trails of radiation-damaged material produced when energetic particles penetrate a solid, such as a mineral grain, are enlarged and made microscopically visible by etching polished surfaces with chemical reagents such as HCl, HF, and KOH. The length of the individual tracks is a measure of the energy of the impacting particles. The track density is an integrated sum of the particle flux and the length of time of irradiation.

From the preceding discussion it is clear that solar wind and solar flare bombardment will affect only the exposed surface of the lunar regolith, whereas cosmic-ray particles may penetrate some centimeters into the surface material. Track studies on Apollo 11 materials have given significant information on the nature of the solar and cosmic-ray irradiation, and on lunar surface processes. For example, it was possible to determine the recent "top" surface of rock 10017 and at the same time to demonstrate that the present "top" and "bottom" must have interchanged in the past, that is, the rock must have been tumbled about on the lunar surface (Fleischer et al., 1970). These studies have also shown that the rate of erosion of rocks on the lunar surface is about 10^{-7}cm/year—it would take 10 million years for 1 cm of material to be removed from the surface of an exposed rock. The abundance of cosmic-ray-produced nuclides shows that some rocks have been on or within several centimeters of the surface for at least 10 million years and within 1 or 2 m for at least 500 million years.

MICROBRECCIAS

The microbreccias (Fig. 2-6) are an abundant and characteristic rock among the Apollo 11 samples. The term *breccia* refers to the abundance of angular rather than rounded grains. The prefix "micro" refers to the very small size of most of the fragments compared to terrestrial breccias. The microbreccias are accumulations of debris most likely generated by impact events. In all, about half of the returned Apollo 11 hand specimens are microbreccia. On Earth breccias are of diverse origin and include rocks fragmented by volcanic explosions or by meteorite impact, and by fracturing in fault zones.

The microbreccias appear to be the lithified equivalent of the lunar regolith. The distribution of grain sizes of the constituent particles is about the same as in the regolith. Furthermore, the mineral and rock types and the abundance of solar wind gases is about the same in each. A principal problem is the determining of the process or processes that led to the lithification of lunar regolith. There are at least two possibilities: (1) shock lithification, and (2) welding in still-hot deposits ejected from impact craters.

Strong shock waves can cause bonding and interlocking of grains as they pass. At peak shock pressures in excess of about 200 kb, fine fractions of loose material melt, causing even more intense lithification (Short, 1970). These shock-lithified rocks are sometimes referred to as "instant rock."

In large impact events, rapidly moving clouds of debris, termed a *base surge,* may be generated. Debris deposited by a base surge may be thick and remain quite hot, even hot enough to cause the welding together of edges of glass particles. Such welded rocks are quite hard, and this mode of lithification is believed by some to be more prevalent in the microbreccias than is shock lithification (McKay et al., 1970).

EVIDENCE OF IMPACT IN THE LUNAR SAMPLES

Intense, impact-generated shock waves have produced marked effects on some of the lunar samples. These effects depend on the material through which the waves pass and on the peak pressure in the wave. Figure 5-4 summarizes these effects for the major lunar minerals. At low to moderate shock pressures, planar features are developed, and some fragmentation occurs. At very high shock pressures, and in some materials excavated by the impact, partial or complete melting may occur. Some of the principles of impact effects in the solid state (impact metamorphism) and in generating glasses have recently been reviewed (French and Short, 1968).

Most mineral and rock fragments in the fines and in the microbreccias do not show evidence of shock metamorphism. However, it is still likely that all the fines are impact ejecta because debris from a single impact event have been studied on Earth and may contain only a few percent of clearly shocked material (Quaide and Oberbeck, 1969). Although all the material is fragmented, only rare ejecta show evidence of solid-state penetrative deformation. Indeed, although low, the percentage of shock-damaged material in the lunar fines and microbreccias appears too high to be attributed to a single impact event. For example, King et al. (1970) observed that of 585 crystalline basalt fragments from the regolith in the

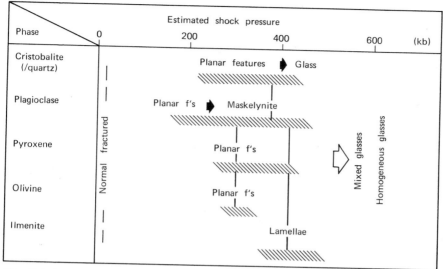

Fig. 5-4. Intense shock waves have marked effects on lunar minerals. At low to moderate pressures, planar features develop. At higher pressures the minerals are converted to glasses. (Douglas et al., 1970.)

size between 1 cm and 1 mm, 35% had evidence of shock damage. Furthermore, the occurrence of fragments of microbreccia included in microbreccia indicates that much of the lunar fines have gone through two or more impact events.

The abundance of impact effects in the lunar regolith and microbreccias is not surprising in view of the ancient age of the crystalline rock basement. Presumably, the surficial deposits have been subjected to bombardment by both meteoritic and secondary impact particles (lunar ejecta) for about 3.7 billion years, which appears to be the solidification age of the basalts in the area.

The Apollo 12 samples also show impact effects. Glasses and deformed silicates are abundant, although very few samples of microbreccia were found. This rarity of microbreccia may be related to the thinner regolith at the Apollo 12 site, which is about half as thick as at Tranquillity Base, and to the possibly younger age of the probable basement rocks.

LUNAR GLASSES

Aside from the small amount of residual glasses in the lunar basalts, all the Apollo 11 glasses have been ascribed to shock-induced melting of prexisting rocks due to meteorite impacts, possibly at peak pressures in the megabar range (Von Engelhardt et al., 1970). At the site of impact, large quantities of melt evidently are ejected as a fine spray, which form droplets with some sort of rotational symmetry (Fig. 5-1). Already solidified or partially solidified glasses become fractured, giving rise to angular fragments (Fig. 5-2). Some glass is envisioned as remaining molten until it collides with rocks, producing the common glass crusts found on both Apollo 11 and Apollo 12 samples.

The rounded small glass bodies are a common and surprising feature of the lunar microbreccias and fines. The bodies may be perfectly spherical, oblate spherules, teardrops, or dumbbell shapes (Fig. 5-1). Their color ranges from dark reddish brown and nearly opaque to colorless. Most are totally glassy, although some show evidence of partial crystallization or incomplete melting.

The chemical composition of the glasses shows that most were produced mainly by the melting of various proportions of the main minerals: pyroxene, plagioclase, ilmenite, and minor olivine (Fig. 5-5). A number of investigators recognized two major compositional classes, which correspond closely to this color and index of refraction: a colorless to light green glass with compositions suggestive of the melting of considerable amounts of very calcic plagioclase, and darker colored glasses richer in iron and titanium. A number of mineral glasses, that is, glasses whose compositions are very close to mineral compositions, were found. These include anorthite, olivine, and pyroxene glasses (Table 5-1). These glasses require very high temperatures for their melting. For example, the anorthite glass spheres require temperatures in excess of 1550°C, the melting point of anorthite at low pressure (1 atm). Such temperatures are far above the liquidus of the lunar basalts and are sufficiently high so that volatilization effects such as the loss of sodium and silicon may be expected. Indeed, such loss is indicated by comparisons of these glasses with likely parent materials (Fredriksson et al., 1970). Table 5-1 gives the ranges in the compositions for various glasses.

Some glass spheres contain a coating of imbedded unmelted grains presumably "captured" by the molten glass in the explosion cloud generated by the impact (McKay et al., 1970). These microscopic particles of molten glass and metal plus crystalline debris evidently underwent numerous collisions with one another in the explosion cloud, as evidenced by the surface features of the debris. Some of the glass spherules have a layered structure, which preserves the sequence of events in the explosion cloud. Evidently, there were first collisions and capture of numerous minute nickel-iron spherules, which typically range from 0.05–0.30 mμ. These metallic spherules and partial crusts (Fig. 5-6) presumably formed either

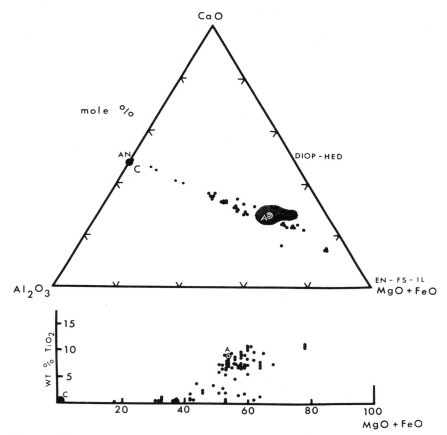

Fig. 5-5. The lunar glasses are derived by impact melting, mainly of various proportions of plagioclase, clinopyroxene, and ilmenite. Each dot represents an electron-probe analysis of a single glass spherule, dumbbell, or angular fragment. The shaded area represents a cluster of 48 analyses. Point A is a scoriaceous glass analyzed by classical wet methods. Point C is the average of 15 plagioclase glasses, which range from 61–94% normative anorthite. In general the density of the glasses increases from left to right, from less than 2.6 to 3.25 g/cm. Abbreviations are: an = anorthite, diop = diopside, hed = hedenbergite, en = enstatite, fs = ferrosilite, il = ilmenite. The bottom figure shows approximate correlation between TiO_2 and FeO plus MgO contents. Lettered points are the same as above. (Frondel et al., 1970.)

by condensation of vaporized nickel-iron or a spray of molten iron from an impacting iron meteorite. Such nickel-iron spherules have been recorded in terrestrial impact-produced glasses (Spencer, 1933). If these metal spherules are condensates, temperatures may have been in excess of 3000°C, the boiling point of iron, at the center of impact. At such high

Table 5-1. Composition of Various Apollo 11 Glasses. Analyses by Electron Probe

	1	2	3	4	5	6	7
SiO_2	46	45.4	46.0	40	41.1	55.4	43.5
TiO_2	0.8	0.1	0.4	9	9.2	0.9	0.4
Al_2O_3	24	34.5	24.6	11	10.3	1.4	6.9
FeO	7	0.3	5.9	19	18.9	10.4	22.1
MnO	0.08		0.1	0.20	0.3		0.3
MgO	8	0.2	8.0	10	6.8	16.7	17.4
CaO	13	19.0	15.1	11	10.5	20.1	8.4
Na_2O	0.6	0.8	0.2	0.5	0.5		0.2
K_2O	0.1	0.1	0.1	0.1	0.2		0.1
Sum:	99.58	100.4	100.4	100.80	97.8		99.3

1 Average of 13 colorless to green glasses. Von Engelhardt et al. (1970) p. 670.
2 Glass of anorthite composition. Index of refraction = 1.5701. Chao et al. (1970), p. 646.
3 Feldspathic glass. Index of refraction = 1.5942. Chao et al. (1970), p. 646.
4 Average composition of 13 yellow, brown, red or violet glasses. Von Engelhardt et al. (1970), p. 670.
5 Titaniferous glass. Index of refraction = 1.7001. Chao et al. (1970), p. 646.
6 Pyroxene glass. Duke et al. (1970), p. 649.
7 Glass rich in olivine component. Chao et al. (1970), p. 646.

temperatures, some direct decomposition of ferrous oxide might occur. This process, presumably, would give low-nickel iron spherules.

McKay et al. (1970) attribute the formation of the glass spherules to at least six possible processes, all related to impact: (1) the expansion and tearing apart of large masses of molten glass formed near the center of major impacts, (2) the breakup of impact-produced liquid jets into droplet trains, (3) the splash and rebound from objects hitting molten masses of glass, (4) the drag of splattered glass over hard surfaces, (5) the condensation of droplets from a vapor cloud, and (6) the vesiculation of impact-produced volcanic glass. They believe that the major lunar process may be production of the spheres by vesiculation, leading to broken walls, which then are pulled together by surface tension into spheres.

There is some evidence that direct condensation of silicates occurs in the impact-generated explosion cloud. McKay et al. (1970) found a smooth coating of high silica, high K_2O glass on the wall of a broken vesicle. Such a composition is believed likely for a silicate condensate.

The enrichment of the heavier isotopes of oxygen and silicon in glasses, microbreccias, and fines compared to the basaltic rocks may be a result of vapor fractionation during boiling in high-temperature impact events (Epstein and Taylor, 1970). Walter and Clayton (1967) found that

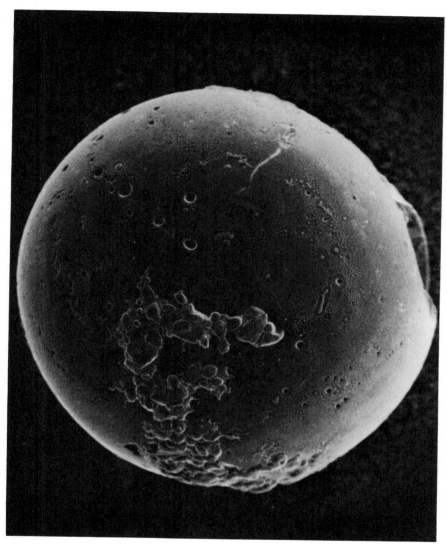

Fig. 5-6. Lunar glass spherules are in some cases partially surfaced by metallic nickel-iron globules. The glass has a composition like the Apollo 11 basalts; the metal globules contain about 3.5–4% nickel, 0.3% phosphorus, and 0.3% cobalt. The spherule is about 0.32 mm in diameter. (Analyses and photographs by B. Glass, NASA Goddard Space Flight Center.)

vapor fractionation does enrich glasses in oxygen 18, but measurements of the silicon isotopes were not made. One might expect that vapor fractionation would have an analogous effect, though, on the silicon isotopes.

Some of the glasses are not homogeneous. Bands and areas of higher index glass are mixed with lower index glass. In some fragments, marked color differences occur between the bands. This heterogeneity arises from melting and incomplete homogenization of coexisting mineral grains. Rarely, crystalline rock is seen which has undergone partial fusion without disruption (Fredriksson et al., 1970). In these, light-colored lower index glass is seen to arise by the melting of plagioclase, whereas darker-colored, reddish orange glass with higher refractive index is associated with the melting of pyroxene and ilmenite.

The average composition of the glass in the lunar fines indicates that its average bulk density is about 3.0 g/cm. However, if one attempts to separate it from accompanying rock and mineral fragments by density flotation, one finds that the individual particles range in density down to 2.5 g/cm, and even less. The reason for this is an extreme vesicularity in some particles down to submicroscopic dimensions, as illustrated in an electron micrograph of a typical glass fragment (Fig. 5-7). At the time of formation of the glass the material must have been thoroughly permeated by a transient gas phase. The vesicularity and extreme freshness of the glass may explain the stimulating effect of the lunar soil on the growth of plants, which has been the subject of considerable comment. Fresh glass surfaces are highly reactive and, on contact with water, readily release inorganic plant nutrients.

GLASS CRUSTS

The large glass crusts that the astronauts observed only within a number of small craters at Tranquillity Base, 1 m in diameter, are of controversial origin. Most scientists have regarded them as simply impact-derived glass spatter. However, this does not readily account for their main occurrence within the small craters. Gold (1970) has proposed an alternative origin for these glass crusts. He envisions localized melting due to radiation heating, perhaps caused by mininova or superflare events on the sun within the last 100,000 years. But why should the areas of glass crusts be restricted to the crater floors? Gold attributes this to some sort of focusing effect of the craters, resulting in concentration of the radiant energy on the crater floors. These large glassy blebs were not collected by the astronauts. Whether or not these are identical to the glassy crusts on the re-

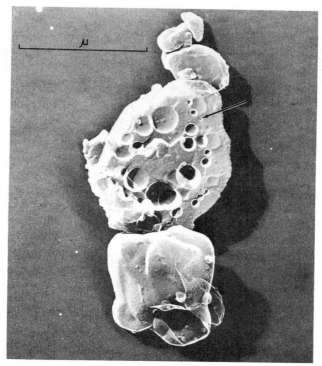

Fig. 5-7. Gas cavities in some lunar glass fragments range down to 150 Å (arrow). Such minute vesicles are extremely rare in terrestrial volcanic glasses. (Electron micrograph by Ken M. Towe, Smithsonian Institution.)

turned rocks and on smaller fragments in the soil, which have already been discussed, is unknown. Some glass crusts on smaller fragments have been regarded as being produced by melting in the hot blast of gases, including volatilized silicates, produced by large impacts.

As at Tranquillity Base, glasses were found to be abundant at the Apollo 12 site in the Oceanus Procellarum. Minute rounded and angular glass particles occur in and on the regolith surface. Also, glass spatter was noted on some of the rocks at the surface and in shallow craters. Detailed examinations of these glasses are not yet available.

MICROCRATERS

Abundant surface microcraters are a characteristic feature of the Apollo 11 samples (Figs. 2-6 and 5-8). These range from the limit of resolution

Fig. 5-8. Pellets of meteoritic nickel-iron are rarely found in the lunar soil. The pellet shown here, collected during the Apollo 11 mission, is marked by hypervelocity microcraters, some of them produced by secondary impacts of lunar material. The pellet is about 4 mm in diameter and weighs 88 mg. Its internal features are shown in Fig. 3-8. (Smithsonian Institution photograph.)

with a binocular microscope to those readily visible in hand specimens; they are believed to have been caused by hypervelocity impacts by micrometeorites or by lunar materials ejected by impacts (secondary impacts).

Microcraters are typically glass-lined and have a raised lip. Outside the lip the craters are surrounded by a whitish highly fractured zone in impacted silicates. Microcraters on one iron-nickel fragment (Fig. 5-8), presumably a fragment of a meteorite, were coated with a glass that has a composition similar to the lunar basalts, suggesting cratering by secondary impacts (Fredriksson et al., 1970). Such microcratering obviously contributes to the disintegration of coarser rock fragments on the lunar surface. However, the density of microcraters is too low to account for all the observed disintegration involved in production of the lunar regolith. Microcratering may occur either in stationary fragments exposed on the lunar surface, or in hypervelocity impacts between large and small fragments within impact explosion clouds.

LUNAR "ANORTHOSITE" AND OTHER MINOR ROCK TYPES

The microbreccias and fines of Apollo 11 contain a minor amount of exotic rock fragments, that is, rock fragments quite unlike the abundant type A and type B fragments. These exotic fragments are believed to represent material from very distant sources, perhaps, as the evidence indicates, from the lunar highlands.

One of the most abundant of these exotic rock types is principally anorthitic plagioclase, along with small amounts of mafic minerals, and rarely with glass. These have been termed *anorthosites* by numerous authors. However, their grain size is much smaller and their plagioclase much more calcic than abundant terrestrial anorthosites. As used for terrestrial plagioclase-rich plutonic rocks, the term "anorthosite" is somewhat misleading, as the abundant pre-Cambrian anorthosite massifs are not composed mainly of anorthite, but of andesine. These lunar fragments will probably continue to be called anorthosites, since they are closer to anorthosites than to any other abundant terrestrial rock.

Chemically, these plagioclase-rich fragments are distinctly different from lunar "basalts" (Table 5-2). The high plagioclase content results in much higher aluminum, calcium, and silica contents. This makes them more comparable to probable lunar highland materials, particularly those at the Surveyor 7 site (compare with Table 1-1). Some of the plagioclase-rich fragments contain up to 50% glass (Mason et al., 1970). In one sample this glass has the normative composition of an olivine-rich basalt. Some fragments appear to have undergone impact metamorphism. (See Fig. 2-6; note colorless large inclusion in the center of the bottom photomicrograph of microbreccia.)

One might speculate that these plagioclase-rich fragments, which are

Table 5-2. Comparison between Lunar "Anorthosites" and Lunar "Basalts" of Apollo 11

	Lunar anorthosite		Av. Apollo 11 basalt
	1	2	3
SiO_2	46.0	45.4	40.38
TiO_2	0.3	tr	10.90
Al_2O_3	27.3	33.8	9.43
Cr_2O_3	0.2	tr	0.33
FeO	6.2	2.8	19.32
MnO	0.1	0.2	0.26
MgO	7.9	1.7	7.20
CaO	14.1	17.5	11.05
Na_2O	0.3	0.4	0.46
K_2O	tr	tr	0.17

¹ Anorthositic gabbro, bulk composition of plagioclase-rich soil fragment determined by averaging the results of ten randomly placed defocused electron beam analyses (Wood et al., 1970). Also report $SO_3 = 0.1$.

² Ibid. Anorthosite.

³ Average of 11 selected analyses of type A and type B igneous rocks, Apollo 11 (column 3, Table 4-1).

⁴ tr = trace.

at the most but a few millimeters across, are shock-melted and recrystallized single crystals of plagioclase, with bits of adhering mafic minerals, derived from the coarser-grained lunar basalts. However, a comparison between the chemical composition of the microbreccias and fines with the average lunar basalt shows that this is unlikely. The fines and microbreccias have higher alumina contents than the basalts; this difference is readily accountable for by the addition of a few percent of these plagioclase-rich rocks, which contain up to 34% alumina (Table 5-2), to the fines and microbreccias.

Other minor rock types include basaltic rocks that have been recrystallized and have been termed *hornfels* (Chao et al., 1970). There are, however, remarkably few accounts of metamorphic rocks that could not be attributed to impact metamorphism. Fragments that are igneous cumulates have also been reported (Chao et al., 1970).

THE METEORITIC INCREMENT

On superficial consideration it might be expected that the lunar surface would be littered with meteorites, accumulated over the many millions of years that most of this surface has been in existence. On Earth freshly fallen meteorites are buried in sediments or rapidly destroyed by weathering, but the absence of normal sedimentary processes and weathering on the Moon should ensure their preservation. However, the situation is not quite as it seems. The absence of an atmosphere means that meteorites hit the lunar surface with unchecked cosmic velocity, and their kinetic energy will be converted to heat sufficient to melt both the meteorite itself and a considerable amount of the impacted rock. Nevertheless, although we may not expect to collect meteorites on the lunar surface, we may hope to recognize a meteoritic increment by its distinctive chemical features.

This has been borne out by the results of the Apollo 11 researches. The breccias and the lunar soil contain pellets and fragments of nickel-iron (Fig. 5-8), whose composition is strongly indicative of a meteoritic origin. The breccias and soil also show notable enrichment in siderophile elements, such as Ni, in comparison to the crystalline rocks; this enrichment is plausibly due to the addition of meteoritic material. The statistics of meteorite falls on Earth indicate that the ratio of stony to iron meteorites is about 9:1, and presumably the proportion among meteorites impacting the Moon is similar. However, the silicate material of the stony meteorites will have been thoroughly blended with the lunar soil at the point of impact (much or all of it fused to glass) and is not recognizable as such. Some microscopic intergrowths of crystals and glass in lunar breccias and soil have been interpreted as fragments of meteoritic chondrules but they may also have formed on the Moon. The amount of recognizable meteoritic nickel-iron in the lunar regolith is quite small, of the order of 0.1%.

Keays et al. (1970) have studied the relative abundances of a number of minor and trace elements in the lunar crystalline rocks, breccias, and soil, and interpret their results as indicating the addition of 1–2% of carbonaceous chondrite material to the lunar surface (Fig. 5-9). Although carbonaceous chondrites are rare among meteorites found on Earth (of about 1000 chondrites in collections, only 30 are carbonaceous chondrites), this rarity may in part be due to their unusual properties: they are rather fragile, break up readily in the atmosphere, and are rapidly destroyed by weathering. Their absolute abundance is probably much greater than the statistics indicate, so that their apparent prominence in the meteoritic increment on the lunar surface is understandable. However, the postulate of a 1–2% addition of carbonaceous chondrite material to the lunar regolith, while plausible, does raise another question: Why are carbon and sulfur not enriched to a corresponding extent? These elements are relatively abundant in the carbonaceous chondrites but are not equally enriched in the lunar soil. Their removal as volatile compounds (such as

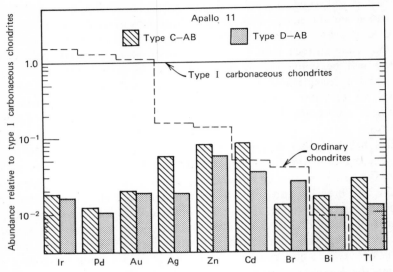

Fig. 5-9. The Apollo 11 fines (D) and microbreccias (C) are enriched in nine trace elements compared to the crystalline igneous rocks (A, B). The "excess" components, obtained by subtracting crystalline rock abundances (A, B) from C and D, show an abundance pattern resembling that of carbonaceous chondrites, not of ordinary chondrites. Apparently the enrichment was caused by the addition of 1–2% of carbonaceous-chondrite like material (Keays et al., 1970).

CO_2, CH_4, H_2S, and SO_2) is a possible explanation, but this problem requires further study.

AGE

The inferred radiometric ages of the fines and microbreccias are an average age for the various crystalline and glassy fragments that compose them. They do not indicate when the fragmentation that produced the fine material occurred. Evidently, most of the fragments composing the fines are very old. Their indicated age of 4.6 billion years by the lead-uranium-thorium and rubidium-strontium methods goes back to the time when the solar system was thought to have formed.

This very old age of the fines and microbreccias presents a problem. Mineralogically, the soil is composed predominantly of rock fragments and mineral fragments like those that would be expected from fragmentation of the large type A and type B samples, which have the much younger age of 3.7 billion years. There has been a definite influx into the regolith of meteorite fragments and fragments probably derived from the lunar highlands. These would indeed tend to raise the age of the fines, but there are only relatively small amounts of these fragments. Thus one might postulate that many of the fragments in the soil, although chemically and mineralogically like the larger samples, are derived from much older rocks.

CHAPTER 6

LUNAR PETROLOGY: COMPARISONS WITH TERRESTRIAL ROCKS, METEORITES, AND TEKTITES

INTRODUCTION

How do the Apollo samples compare with terrestrial rocks and meteorites? In many ways they are quite similar. No new elements have yet been found, and most lunar minerals also occur on Earth. Also, the ratios of stable isotopes are mainly the same. On the other hand, there are a number of very important differences. The crystalline igneous rocks from Apollo 11 are older than any known terrestrial rocks, and contain a number of minor minerals that either are very rare or have not yet been found in terrestrial rocks. Also, the evidence of impact events is much more prevalent in the lunar samples than in terrestrial samples. The following sections deal primarily with comparisons with terrestrial and extraterrestrial materials which have at least some similarity to the lunar rocks or have been speculated to have come from the Moon (tektites and certain meteorites).

LUNAR AND TERRESTRIAL BASALTS

Of the four major Apollo 11 rock types, only two have abundant terrestrial counterparts: the type A and type B crystalline igneous rocks. Even these differ, however, in fundamental ways from their closest terrestrial equivalents. The lunar microbreccias and fines have no abundant counterpart among terrestrial rocks. If large meteorite impacts were common in the Earth's early history, such rocks would have been abundant. However, erosion would have long since erased most evidence of these deposits.

Only in localized areas around recent and large ancient meteorite impact sites are similar rocks found on Earth; terrestrial impact breccias are known as *suevites*.

The Apollo 11 crystalline rocks have been termed *basalts* by numerous authors. This is not unreasonable usage of the term "basalt" because these rocks are closer to basalts than to any other major terrestrial igneous rock group. Like terrestrial basalts the lunar crystalline rocks consist mainly of augite and plagioclase and either cooled on or very near the lunar surface; that is, they are volcanic rocks.

A number of more specific terms have been given to the lunar basalt to reflect certain important differences compared to abundant varieties of terrestrial basalts. Anderson et al. (1970) termed them *ferrobasalts*, to indicate their unsually high ferrous iron content. This high FeO content, like the high titanium content, is due to the abundance of ilmenite. The subtraction of most of the ilmenite (Table 6-1) makes the major and minor element composition somewhat closer to an average ocean-ridge basalt, one of the most abundant types of terrestrial basalts. However, fundamental differences remain. Following the detailed igneous rock classification of Johannsen (1937), which is based on mineral abundances, the lunar basalts have been termed *ilmenite melabasalts* and *ilmenite meladolerites*. The differences between mineral abundances in the lunar

Table 6-1. Composition of Average Apollo 11 Igneous Rock Recalculated Minus Ilmenite Compared to an Average Deep-Sea Basalt and a Titanium-Rich Deep-Sea Gabbro

	1	2	3	4
SiO_2	40.38	49.71	49.21	41.83
Al_2O_3	9.43	11.61	15.81	11.94
FeO	19.32	13.62	8.18	18.39
MgO	7.20	8.86	8.53	6.47
CaO	11.05	13.60	11.14	9.88
Na_2O	0.46	0.57	2.71	2.43
K_2O	0.17	0.21	0.26	0.10
MnO	0.26	0.32	0.16	0.21
TiO_2	10.90	1.71	1.39	7.05
P_2O_5	0.12	0.15	0.15	0.01
	99.29	100.36	97.54	98.31

[1] Average of 11 type A and type B crystalline rocks (analysis 3, Table 4-1).

[2] Column 1 recalculated to 100% after subtracting 18% normative ilmenite.

[3] Average ocean-ridge basalt (Melson et al., 1968a).

[4] Clinopyroxene–hornblende–titanomagnetite gabbro (Miyashiro, Shido, and Ewing, 1970, p. 363). Contains 7.05 Fe_2O_3 which was recalculated as FeO.

102   THE LUNAR ROCKS

and some common terrestrial basalts (flood basalts, and basalts from the midoceanic ridges) are shown in Fig. 6-1.

As on Earth, the lunar basaltic rocks appear to be most abundant in great basins, the maria. Basalts are the dominant volcanic rock of the ocean floors on Earth. They also occur on the continents, where they commonly form great flood basalt fields such as the Columbia River Plateau. However, most basalts appear to be in the oceanic crust. Also, grain-size variations between the fine and coarse-grained lunar samples are not unlike those found in thick flood basalt flows (Fig. 6-2).

Also, the lunar basalts, like terrestrial flood basalts, evidently do not build marked volcanic edifices around their vents. This is largely a result of their low viscosity. Lunar basalts from Apollo 11 have even lower viscosities than terrestrial flood basalts.

The high content of ilmenite is not a unique feature of the Apollo 11

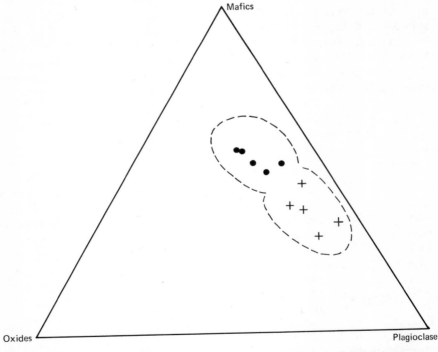

Fig. 6-1. The lunar "basalts" from Apollo 11 (solid circles) are richer in opaque minerals (mainly ilmenite) and in pyroxene than abundant varieties of terrestrial basalts (crosses). The terrestrial basalts shown include deep-sea basalts and basalts from the Columbia River Plateau flood basalts. Mafics refer to olivine plus pyroxene but are mainly pyroxene. Plotted in vol %. (Mason and Melson, 1970.)

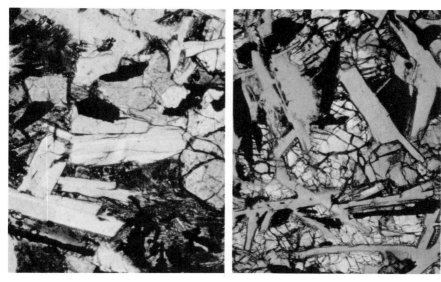

Fig. 6-2. Samples from the interior of thick terrestrial flood basalt flows (left) are coarser-grained than some of the type B lunar basalts (right). Terrestrial samples are from the Yakima basalt, Washington (USNM 109406, 3). The lunar sample is a fragment from the fines (10084). The width of the field is about 0.5 mm in both graphs. Note the greater abundance of interstitial fine-grained material (mesostasis) in the terrestrial sample.

basalts. Dark-colored rocks even higher in ilmenite have been reported from the Skaergaard intrusion, east Greenland (Brown et al., 1970). However, these rocks were enriched in ilmenite by crystal settling in a large intrusion and are much higher in potassium and sodium (Brown et al., 1970). Ilmenite-rich rocks also are found in terrestrial Precambrian, gabbro-anorthosite complexes. However, these also differ markedly in other regards from the lunar samples. They are plutonic rocks, that is, they cooled at considerable depth within the Earth and rarely contain anorthitic plagioclase. The processes often invoked to explain these terrestrial ilmenite-rich rocks—fractional crystallization, oxide-liquid immiscibility, and gravitative settling of ilmenite—do not account for the high ilmenite content of lunar basalts (Cameron, 1970).

High-titanium basaltic rocks have been dredged from the deep-sea floor of the mid-Atlantic Ridge. These rocks, which have been termed gabbros and are described by Miyashiro et al. (1970), were shown to have titanium and total iron contents close to those of the Apollo 11 basaltic rocks (Table 6-1). On the other hand, they are much higher in sodium and potassium, and will undoubtedly have certain distinctly different trace element contents. Contrary to the lunar samples, brown hornblende appears to be an important host mineral for titanium. Also, the reported oxide mineral is titanmagnetite rather than ilmenite and the rock is high in ferric iron. Other terrestrial occurrences of high-titanium basic to ultrabasic rocks include the brown-hornblende mylonitized gabbros of St. Pauls Rocks on the mid-Atlantic Ridge (Melson et al., 1968a). These terrestrial high-titanium basic and ultrabasic rocks are rare. They are nonetheless important, because the as yet unspecified process which led to their great enrichment in titanium may be the same process which produced the high titanium in the Apollo 11 rocks.

The well-developed ophitic texture of the coarse-grained lunar rocks is a characteristic feature of terrestrial basaltic intrusive rocks (Fig. 6-3). However, the plagioclase in the pyroxene-plagioclase intergrowth is much more calcic (bytownite to anorthite) than in common terrestrial basalts (labradorite). Also, the new lunar pyroxenoid, pyroxferroite, found marginal to, and in, the interstices of the coarse-grained lunar rocks has not yet been found in terrestrial basalts. Its presence in lunar basalts is not fully understood, as it appears to have no stability field at low pressures (Lindsley and Munoz, 1969; Lindsley and Burnham, 1970). Metastability, aided by rapid crystallization in an anhydrous magma, may be an important factor in its presence in lunar rocks and absence in terrestrial rocks (Fig. 6-4).

Metallic iron, found in both Apollo 11 and 12 igneous rocks, does not normally occur in terrestrial basalts; in the lunar basalts, it evidently separated from an immiscible sulfur-iron melt (Skinner, 1970). Terrestrial basalts crystallize with relatively higher oxygen-metal ratios (that is, they are more oxidized) than the lunar basalts. Indeed, Anderson et al. (1970) postulate that the oxygen fugacities—a measure of oxygen availability during crystallization—was a thousand times less than that of common terrestrial basalts. Metallic iron does occur rarely in terrestrial basalts. The most notable occurrence is at Disko Island, west Greenland, where the metal was most likely produced by the reduction of basaltic magma contaminated by carbonaceous shale inclusions (Melson and Switzer, 1966). Like the lunar rocks, and unlike most other terrestrial basalts, the principal oxide mineral is ilmenite. In other regards, the Disko basalts are closely akin to other terrestrial basalts. They have relatively high alkali and low titanium contents, and the plagioclase is labradorite.

Certain features of the lunar basalts resemble those of the terrestrial alkali basalt family. These do not include two of the most important ones: the high alkali content and presence of normative nepheline, two of the

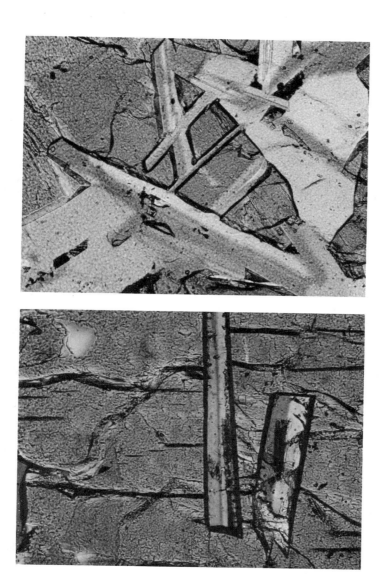

Fig. 6-3. Augite crystals (gray), enclosing plagioclase crystals (lighter colored, elongate), is a common texture (termed ophitic) in both lunar (top) and terrestrial (bottom) basaltic rocks. The lunar sample is Apollo 11, type B rock 10047; the terrestrial sample is a dolerite (USNM 111889).

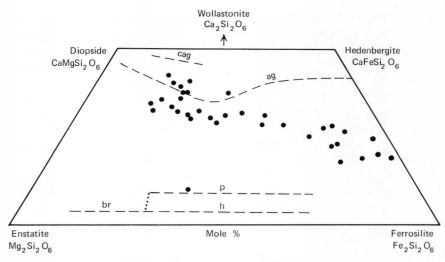

Fig. 6-4. Terrestrial and lunar pyroxene composition trends differ markedly. Terrestrial pyroxenes are indicated by broken lines, where the abbreviations are cag = calcic augite; ag = augite; br = bronzite; h = hypersthene; and p = pigeonite series. The lunar pyroxenes (solid circles) are from three crystalline rocks. (Modified from Brown et al., 1970.)

features that distinguish this family. However, as in alkali basalts, the olivine, near Fa_{30}, may contain octahedra of a brown spinel (picotite, Agrell et al., 1970). Like certain basanites, olivine-rich undersaturated terrestrial basalts, olivine in the lunar basalts is frequently rimmed by clinopyroxene. The alkali basalts, like the lunar basalts, contain very little glass, even when cooled very rapidly. In these and in the lunar samples, this feature is related to the low silica contents and hence low viscosity.

The presence of residual liquids of granitic compositions (p. 68) is not a unique feature of the lunar igneous rocks. Residual glasses in certain Hawaiian tholeiitic basalts are also granitic (Evans and Moore, 1968). Also, like the Apollo 11 basalts, most abundant terrestrial basalts, from both the sea-floor and the continents, are close to saturation or just oversaturated with regard to silica.

Another important similarity is the solidus and liquidus temperatures (Fig. 4-7). The Apollo 11 basalts and common terrestrial basalts are both totally molten around 1150°C, and have a relatively small crystallization interval. This is evidence that both are the products of small amounts of partial melting.

The Apollo 12 igneous rocks are more variable. These variations make some samples closer to terrestrial basalts than the Apollo 11 samples.

This includes the lower titanium content of most. However, pigeonite, evidently a common phenocryst in some Apollo 12 samples, is rare as phenocrysts in terrestrial lavas. When found as such, it occurs in dacites or andesites. In terrestrial basalts its principal occurrence is as a groundmass mineral, as it is in most Apollo 11 samples.

In summary, lunar and terrestrial basaltic rocks have numerous similarities and dissimilarities. Perhaps some of their most important differences are in certain minor and trace elements, as discussed in the following chapter. Important genetic similarities include: (1) the apparent abundance of both in basins, the maria on the Moon and ocean basins on Earth; and (2) their similar melting temperatures and melting intervals. We might conclude that both lunar and terrestrial basalts are products of partial fusion and may be erupted as flood basalts of low viscosity but that they are derived from compositionally different parent materials.

The assumption that the lunar magmas are partial melts permits inferences about the composition of their source region. Ringwood and Essene (1970) have compared the probable properties of the source region for the Apollo 11 basaltic rocks with the source region for terrestrial ocean-ridge basalts (Table 6-2). This comparison shows much higher Fe/Fe +

Table 6-2. Comparisons of Composition of Probable Source Regions for Terrestrial Oceanic Basalts and Apollo 11 Basalts (Ringwood and Essene, 1970)

Parameter	Earth	Moon
Chemical composition	Dominantly Mg-Fe silicates in both, with less than 5 weight % each of Al_2O_3 and CaO	
Mineralogy	Dominantly olivine	Dominantly pyroxene
FeO/(FeO + MgO) in mole %	12	25
Oxygen fugacity at 1200°C	10^{-8}–10^{-9}	$10^{-13.5}$
Siderophile elements Ni, Cu, Ga	Apparently depleted in Moon as compared to Earth	
Fractionation of volatile elements* Rb, K, Na, Zn, and Pb		
Rb/Ba	0.10	0.03
K/Ba	93	11
K/U	10,000	3000
Na/Al	0.22	0.06
Zn/Mn	0.07	0.01
Pb/Ba	0.06	0.02

* To offset the effects of crystal-liquid fractionation upon absolute abundances, comparison is made on the basis of abundance ratios in which the abundance of each volatile element is compared to the abundance of a relatively nonvolatile element which behaves similarly during crystal-liquid fractionation processes in relevant silicate systems.

Mg ratio for the lunar interior. It also shows the greater depletion of the "volatile" elements in the lunar interior. Other differences include the lower oxidation state and depletion of nickel, copper, and gallium in the lunar interior.

METEORITES AND THE LUNAR ROCKS

One of the early theories for the origin of meteorites was that they are rocks ejected by volcanic eruptions on the moon. This theory was abandoned when it was realized that the energy of volcanic eruptions, while great, does not impart sufficient velocity to the ejecta for their escape even from the weaker gravitational field of the Moon. However, the idea was revived in connection with the impact theory of origin for the large craters on the Moon; the numerous secondary craters around these large craters and the extended rays associated with some of them indicate the large amount of material ejected from them. Is it not possible that some of this material was ejected with velocities greater than the escape velocity for the Moon (2.4 km/sec) and eventually landed on Earth? Admitting this possibility, can we identify this material in our meteorite collections?

Many types of meteorite have been suggested for this distinction. (Tektites, which are not universally accepted as meteorites, will be discussed in the following section.) Urey (1959, 1967) has proposed the Moon as a source of the chondrites. He develops the hypothesis that chondrules are produced by melting associated with extraterrestrial collisions and that these collisions occurred on the surfaces of primary objects in the solar system, of which the Moon was one. Wänke (1966) has argued specifically for the bronzite chondrites as lunar-derived meteorites, on the basis of an analysis of their cosmic-ray exposure ages and other related features. Duke and Silver (1967) suggested that the calcium-rich achondrites (eucrites and howardites) and the mesosiderites are lunar rocks. They proposed that the howardites, which are extremely brecciated, are derived from the lunar highlands, while the eucrites, in composition more like basalts than the howardites and usually less brecciated, are derived from the maria. The mesosiderites are essentially mixtures of howardite silicates with nickel-iron, but Duke and Silver do not ascribe them to any specific region of the Moon.

The possibility of material being ejected into Earth-crossing orbit from the Moon by meteoritic or cometary impact is still somewhat speculative. However, even if we accept the possibility, it seems unlikely that the chondrites are derived therefrom. So far, there is no evidence for chondrite compositions on the surface of the Moon; for one element alone, aluminum, the amounts in all rocks so far analyzed are several times greater than in any chondrite. Densities of the common chondrites range from 3.5–3.8, greater than the bulk density of the Moon. The glass spherules in lunar soils and breccias superficially resemble chondrules in size and shape but are quite different in composition.

The case for eucrites and howardites is considerably stronger. The initial analysis at the Surveyor V site (Turkevich et al., 1967) indicated a comparability with calcium-rich achondrites, although the refinement of the data (Turkevich et al., 1969) showed that the titanium content was much higher than in any known achondrite (Fig. 1-7). However, it now appears that this very high titanium content may be specific to Mare Tranquillitatis. Apart from their titanium content the Apollo 11, and even more the Apollo 12, rocks are quite comparable in chemical and mineralogical composition to some of the eucrites. This comparison, for the Moore County eucrite, is illustrated in Table 6-3 and Fig. 6-5.

Table 6-3 shows that subtraction of 20% (by weight) of $FeTiO_3$ from a typical Apollo 11 basalt produces a composition nearly identical with that of the Moore County eucrite; the most marked discrepancy is in the considerably higher K_2O content of the lunar rock. A comparison of thin sections (Fig. 6-5) shows a close similarity in mineralogy and texture. In both rocks, pyroxene and calcic plagioclase are major minerals; the lunar rock, of course, contains considerable amounts of ilmenite, which is an accessory mineral in Moore County. Both rocks are slightly oversaturated with respect to SiO_2, resulting in the presence of accessory tridy-

Table 6-3. Chemical Composition of Lunar Basalt, Compared with the Moore County and Kapoeta Meteorites

	1	2	3	4
SiO_2	39.79	49.31	48.16	48.47
TiO_2	11.44	1.12	0.32	0.37
Al_2O_3	10.84	13.41	15.57	9.46
FeO	19.35	12.26	15.69	17.16
MnO	0.20	0.25	0.31	0.53
MgO	7.65	9.48	8.41	12.00
CaO	10.08	12.49	11.08	8.08
Na_2O	0.54	0.67	0.45	0.46
K_2O	0.32	0.40	0.09	0.05
P_2O_5	0.17	0.21	...	0.07
Cr_2O_3	0.32	0.40	0.44	0.63

[1] Lunar basalt 10057 (Engel and Engel, 1970).
[2] Lunar basalt, recalculated to 100 after subtracting 20% ilmenite.
[3] Moore County eucrite (Hess and Henderson, 1949).
[4] Kapoeta howardite (Mason and Wiik, 1966).

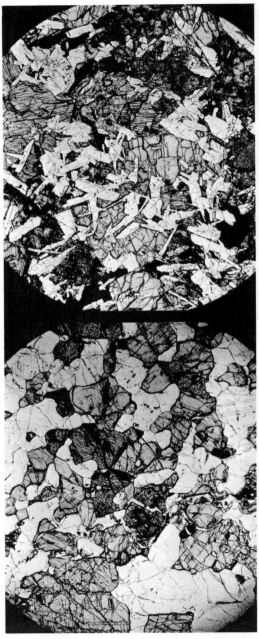

Fig. 6-5. Photomicrographs of thin sections of lunar rock 10044 (top) and the Moore County meteorite (bottom). White is plagioclase with minor tridymite and/or cristobalite; gray is pyroxene, black opaques; magnification 20×.

mite in Moore County, and a mixture of tridymite and cristobalite in the lunar rock. As does the lunar rock the meteorite contains trace amounts of troilite and metallic iron, indicating crystallization at low oxygen fugacity. The pyroxene in the meteorite has undergone a complex series of exsolution and inversion, indicated by lamellae of augite in host pigeonite, partial inversion of pigeonite to hypersthene, and further exsolution of augite from the hypersthene. The pyroxene in the lunar rock generally appears optically homogeneous, but x-ray examination shows that submicroscopic exsolution is certainly present.

The chemical, mineralogical, and textural similarities between the Moore County meteorite and some of the lunar rocks probably indicate a similar origin. Admittedly, the high titanium content of the Apollo 11 rocks is an important difference, but this may be a local feature of these rocks, since the alpha-scattering analyses of Surveyor 6 and 7 and preliminary reports on the Apollo 12 rocks indicate much lower titanium values at these sites. Two deductions seem reasonable. One is the possibility of finding on the Moon a rock mass from which the Moore County meteorite could have been derived. The other is that igneous processes similar to those once active on the Moon have also been active in other regions of the solar system, presumably on the larger asteroids, and have given rise to similar rocks. As was pointed out some years ago (Mason, 1962), the eucrites show chemical and mineralogical features indicating that they represent residual liquids from the fractional crystallization and differentiation of an original melt of average chondritic composition or partial melts from such a composition.

The howardites, exemplified by Kapoeta, resemble the eucrites in composition (except that they consistently contain less plagioclase and more pyroxene) but are characterized by a microbreccia structure (Fig. 6-6), with angular fragments of pyroxene and plagioclase, and occasional composite fragments, in a groundmass of comminuted pyroxene and plagioclase. This structure is similar to that of the lunar microbreccias, and the similarity is enhanced by the presence of rare fragments of nickel-iron in both. The principal distinction is that Kapoeta (and the other howardites, as far as we have examined them) does not contain the glass spherules and fragments that make up a minor part of the lunar microbreccias. Additionally, the mineral grains in the howardites seem to be less shocked than some of those in the lunar microbreccias; in the latter many of the plagioclase grains are partly or completely converted to maskelynite, a feature that appears to be absent in the howardites.

This comparison therefore suggests that the howardites and the lunar microbreccias are the products of a unique process acting on materials of similar composition and structure. Since the lunar microbreccias can plausibly be interpreted as "instant-rock" produced through lithification of the

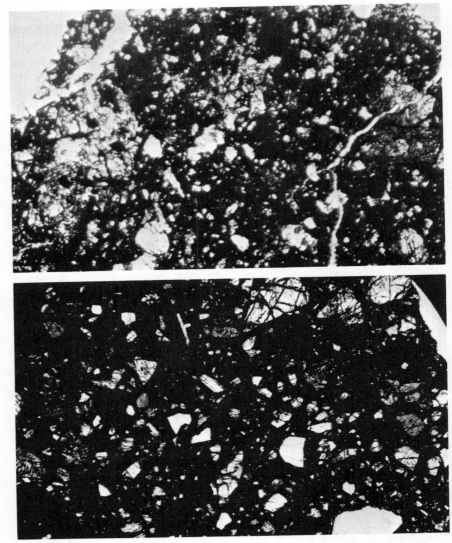

Fig. 6-6. Photomicrographs of thin sections of lunar breccia 10048 (top) and the Kapoeta meteorite (bottom), showing fragments of plagioclase and pyroxene in a semiopaque groundmass of comminuted minerals; magnification 33×.

lunar regolith by meteoritic impact (or perhaps impact by fragments projected from other regions of the Moon), it seems reasonable to postulate a similar origin for the howardites. However, the absence of glass and highly shocked mineral grains in the howardites suggests that the impacts which consolidated them were less violent than those on the Moon, a possible indication of smaller parent bodies and smaller impacting objects, such as might be expected in the asteroidal belt.

Onuma et al. (1970) have observed that the isotopic composition of oxygen in lunar pyroxenes is similar to that in terrestrial pyroxenes but different from that in the calcium-rich achondrites and the mesosiderites; this suggests that these meteorites were not derived from the Moon.

Kopal and Rackham (1964) observed a red luminescence excited by solar activity around the crater Kepler. Derham and Geake (1964) had reported that enstatite achondrites show a similar luminescence, and Kopal and Rackham therefore proposed that the Kepler crater was formed by the impact of an enstatite achondrite. Geake (1964) pointed out that the throw-out from a crater contains only a minute amount of the impacting body, probably insufficient to account for the observed luminescence, and suggested the alternative hypothesis that the lunar material exposed by the impact had the composition of enstatite achondrites. This raises the corollary that the enstatite achondrites may have come from the Moon. However, none of the recent observations support this corollary. Enstatite achondrites have crystallized under such highly reducing conditions that all iron is reduced to the metallic state, whereas all the lunar rocks so far found contain considerable amounts of ferrous iron. None is any way comparable with the enstatite achondrites in either chemical or mineralogical composition.

Tektites and the Moon

Tektites are small pieces of silica-rich (usually 65–80% SiO_2) glass found in limited areas in a few regions on the Earth's surface (Fig. 6-7), under conditions that preclude a volcanic origin. F. E. Suess, who introduced the name tektite in 1900, considered them to be glassy meteorites; however, unlike other meteorite types, tektites have not been seen to fall. Geological and radioactive dating indicates that different geographic groups have different ages: North American tektites, 33 million years; Czechoslovakian tektites, 15 million years; Ivory Coast tektites, about 1 million years; tektites from southeast Asia and Australia, 700,000 years or less. In 1967 microscopic glass spheres similar in composition to tektites were discovered in deep-sea sediments in the Indian Ocean west of Australia and named *microtektites;* similar microtektites have been found in deep-sea sediments off the Ivory Coast.

Two theories are current as to the origin of tektites. Both ascribe the production of the silicate melt to the impact of comets or large meteorites but differ as to the location of the impact: Earth or Moon? If they are

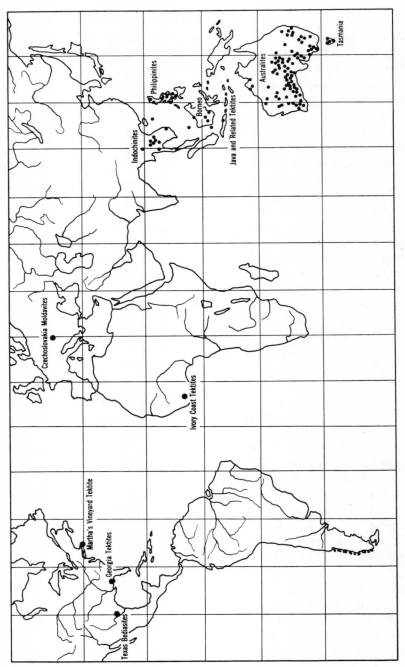

Fig. 6-7. The geographical distribution of tektites. (After Barnes, 1961.)

extraterrestrial in origin, the Moon is the only feasible source of tektites, since their lack of cosmic-ray produced radioactivities implies a brief extraterrestrial passage. Scientists who have worked on the tektite problem have been fairly evenly split between a terrestrial and a lunar origin for these enigmatic objects. For this controversy, therefore, lunar exploration has a special significance.

The evidence from the lunar samples has been generally unfavorable for the origin of tektites from the Moon. None of the lunar compositions from the Surveyor analyses or of the Apollo 11 rocks is within the range of tektite compositions; however, some microtektites are much lower in SiO_2 than the tektites themselves and have compositions comparable to some lunar glasses. Analyses of residual glasses in the Apollo 11 rocks (Table 4-3) show SiO_2 contents in the range of tektite compositions, but these residual glasses are notably lower in MgO and higher in K_2O and are thus significantly different. Nevertheless, among the Apollo 12 collections is one small rock with a composition far removed from the general run of lunar basaltic rocks. This is 12013, and its preliminary analysis (LSPET, 1970) is given here (A), along with that of a Java tektite (B) analyzed by Cassidy et al. (1969):

	SiO_2	TiO_2	Al_2O_3	FeO	MgO	CaO	Na_2O	K_2O
A	61	1.2	12	10	6.0	6.3	0.69	2.0
B	63.5	0.8	12.6	8.5	6.8	3.8	0.7	1.5

The similarities are striking, as pointed out by O'Keefe (1970); they indicate that a possible lunar origin of tektites cannot be ruled out on composition alone (although most tektites contain considerably more SiO_2 than this Java specimen). In addition, tektites, like the lunar rocks, have exceedingly low H_2O contents.

CHAPTER 7

LUNAR GEOCHEMISTRY

One of the principal objectives in the investigation of the lunar samples has been the determination of their composition, both for the major elements and for those present in minor and trace amounts. Not only the amounts of the individual elements, but also, for some of them, the isotopic composition is highly significant for many aspects of lunar history and evolution. Thus a considerable part of the total research effort on the returned samples has been devoted to chemical and isotopic analyses, using a variety of highly sophisticated techniques such as neutron activation, isotope dilution, electron-beam microprobe, spark-source mass spectrometry, emission spectrography, and others. As a result, a vast amount of analytical data has already been published. The coverage is somewhat uneven. For some elements, particularly those of high geochemical significance, such as the radioactive elements and their daughter products, the rare earths, and carbon, the data are very extensive. For others, usually those present in low concentrations for which sensitive analytical techniques are not readily available, the data are still sparse. Nevertheless, the overall coverage is certainly uniquely comprehensive, comparable only with two terrestrial rocks used for many years as geochemical standards: the granite G-1 and the diabase W-1.

The data will be discussed by individual element or group of elements, but in order to provide a rapid review we have compressed these data into the form of Table 7-1. In establishing the range for each element in the lunar materials, we have used our best judgment in eliminating some results that appear erroneous, either by comparison with other analyses on the same sample or by being seriously inconsistent with other data for the specific element. The figure for the Apollo 11 mean should be viewed critically. For those elements with a limited range it has a certain utility, but for those elements for which the range is great it probably has little real

Table 7-1. Abundances (in ppm unless otherwise indicated) of the Elements in Apollo 11 Materials and in Apollo 12 Fines Compared to Terrestrial Basaltic Rock W-1, Eucrites, and Type I Carbonaceous Chondrites. (If no mean is given for the Apollo 11 materials, the data are inadequate for this purpose)

Element	Apollo 11 Range	Apollo 11 Mean	Apollo 12 fines	W-1	Eucrites	Carbonaceous chondrites
Li	9–23	12	11	12	8	1.3
Be	1–6	2	—	0.8	0.1	0.04
B	1–4	2	—	15	0.8	5
C	64–230	140	110	—	—	3.5%
N	30–150	100	—	14	30	2600
O	37.6%–43.4%	40.0%	—	44.6%	42.7%	45.3%
F	30–340	140	—	250	60	190
Na	2600–4000	3300	3000	1.6%	3000	5500
Mg	3.4%–5.1%	4.5%	7.2%	4.0%	4.3%	9.6%
Al	3.7%–7.8%	5.6%	7.4%	7.9%	6.5%	8500
Si	17.7%–20.6%	19.2%	19.6%	24.9%	22.8%	10.3%
P	200–900	500	—	610	400	1400
S	1200–2400	1700	—	130	900	6.2%
Cl	3–30	14	—	200	20	260
K	400–2800	1400	1500	5300	400	1400
Ca	7.2%–9.0%	8.0%	7.1%	7.8%	7.7%	1.1%
Sc	60–100	75	47	34	35	5
Ti	4.3%–7.4%	5.9%	1.9%	6400	4600	420
V	20–100	50	64	240	75	57
Cr	1300–2800	2100	2800	120	2100	2200
Mn	1500–2400	1900	1900	1300	3900	1700
Fe	11.8%–15.6%	14.3%	13.2%	7.7%	14.5%	18.4%
Co	11–35	25	42	50	4	480

Table 7-1. (*Continued*)

Element	Apollo 11 Range	Apollo 11 Mean	Apollo 12 fines	W-1	Eucrites	Carbonaceous chondrites
Ni	3–280	—	200	78	13	1.0%
Cu	4–25	11	—	110	7	140
Zn	2–40	15	5.4	82	2	320
Ga	3–6	4.5	4.9	16	2	10
Ge	<0.1–0.4	—	—	1.7	0.1	34
As	0.01–0.09	0.05	—	2.4	0.05	2.0
Se	0.4–1.6	0.8	0.24	0.11	0.002	27
Br	0.01–0.4	0.1	0.13	0.4	0.4	5
Rb	0.5–6	3.4	8.7	22	0.35	2.3
Sr	110–220	170	170	180	85	8
Y	70–170	120	130	25	23	1.6
Zr	180–660	370	670	100	46	9
Nb	14–31	21	—	10	—	0.5
Mo	0.4–0.7	0.5	—	0.5	—	1.6
Ru	—	—	—	<0.4	—	0.7
Rh	—	—	—	<0.005	—	0.2
Pd	0.001–0.013	0.006	—	0.01	—	0.6
Ag	0.001–0.024	0.008	0.005	0.05	0.04	0.4
Cd	0.003–0.11	0.004	0.05	0.3	0.001	1.0
In	0.003–0.05	0.003	0.009	0.07	—	0.09
Sn	0.3–1.2	0.6	—	3	0.01	1.6
Sb	0.005–0.01	0.007	—	1.1	0.0002	0.15
Te	0.008–0.073	0.02	0.075	<0.2	0.2	3.3
I	0.006–1.4	—	—	<0.05	—	0.3
Cs	0.02–0.17	0.10	0.32	1.0	0.01	0.19
Ba	70–340	200	420	180	35	4
La	7–29	18	—	12	3.7	0.19

Table 7-1. (Continued)

Element	Apollo 11 Range	Apollo 11 Mean	Apollo 12 fines	W-1	Eucrites	Carbonaceous chondrites
Ce	23–83	54	—	23	9.7	0.63
Pr	5–16	11	—	4	1.4	0.09
Nd	21–69	46	—	17	6.9	0.42
Sm	8–23	15	—	4	2.3	0.13
Eu	1.5–2.7	1.9	—	1.1	0.72	0.05
Gd	12–29	20	—	4	2.9	0.24
Tb	2.1–5.0	3.6	—	0.8	0.57	0.04
Dy	14–36	25	—	4	3.8	0.22
Ho	2.2–8.7	4.9	—	1	0.80	0.06
Er	9–21	14	—	3	2.3	0.14
Tm	1.2–2.8	1.9	—	0.35	0.38	0.02
Yb	8–20	13	—	2.2	1.9	0.13
Lu	1.2–2.9	1.9	—	0.35	0.38	0.02
Hf	7–18	13	—	2	0.8	0.32
Ta	1.0–2.7	1.7	—	0.7	0.1	0.02
W	0.1–0.4	0.3	—	0.45	—	0.14
Re	0.01	—	—	0.0003	0.00005	0.04
Os	0.0003	—	—	0.0003	0.0005	0.45
Ir	0.00001–0.01	0.00007	0.009	0.0003	0.0002	0.40
Pt	—	—	—	0.02	—	0.90
Au	0.00002–0.004	0.00004	0.002	0.005	0.001	0.18
Hg	0.0006–0.013	—	—	0.1	—	1?
Tl	0.0003–0.003	0.0006	0.002	0.13	0.0007	0.14
Pb	0.3–1.8	1.2	—	8	0.5	2.9
Bi	0.0001–0.003	0.0003	0.002	—	—	0.13
Th	0.5–3.4	2.0	6.0	2.4	0.4	0.04
U	0.16–0.9	0.5	1.5	0.5	0.1	0.01

119

meaning, although it may be useful for comparative purposes. An additional factor to be considered is that the abundances of some elements are clearly quantized within specific rock types, so that averages for these specific rock types may be more significant than the overall mean. This is particularly true for the breccias (type C) and soil or fines (type D) versus the crystalline rocks (types A and B). The former contain a small but geochemically significant meteoritic increment, which results in markedly higher contents of nickel and geochemically related elements. More subtle differences can be discerned between the individual types. For instance, the Apollo 11 crystalline rocks can be divided into two groups by consistent differences in the concentrations of some of the elements, such as rubidium, barium, the rare earths, thorium, and uranium (Table 7-2). These two groups roughly parallel the textural division of these rocks into type A (fine-grained) and type B (medium-grained), type A rocks showing higher concentrations (two to six times) of these elements than type B rocks. Conversely, other minor and trace elements show essentially uniform concentrations in all the crystalline rocks. One controlling factor is whether a specific element is contained in a major mineral (in which case its concentration is fairly uniform in all the rocks) or is present in an accessory mineral, whose amount may vary considerably in different rocks. The fine-

Table 7-2. Minor and Trace Elements in Apollo 11 Crystalline Rocks, Showing Quantization between Fine-grained (A) and Medium-grained (B) Types

Rock Number	K	Rb	Zr	Ba	Ce	Th	U	Type
17	2610	5.6	476	308	77	3.4	0.85	A
22	2290	5.6	450	277	75	—	0.80	A
24	2400	6.0	375	310	108	4.1	—	B
49	2730	6.2	—	338	84	—	0.74	A
57	2100	5.2	635	280	75	3.4	0.87	A
69	2300	5.6	566	300	65	—	0.78	A
71	2770	5.9	644	327	84	3.4	0.87	A
72	2300	5.7	497	300	80	3.3	0.88	A
03	470	0.5	340	108	45	1.0	0.27	B
20	490	0.6	360	80	26	0.7	0.20	A
44	820	1.2	280	95	42	1.0	0.33	B
45	420	0.6	194	90	23	0.9	0.26	B
47	900	1.2	334	88	48	0.6	0.26	B
50	530	0.7	—	60	34	0.5	0.16	B
58	880	1.0	380	120	40	1.1	0.20	B
62	630	0.9	319	80	43	0.9	0.27	B

grained crystalline rocks contain more fine-grained mesostasis than the medium-grained rocks, and this mesostasis is the seat of accessory minerals (such as apatite and potash feldspar) enriched in minor and trace elements.

In order to utilize the compositional data most effectively for the elucidation of the geochemical evolution of the lunar rocks, it is necessary to compare and contrast the figures for the individual elements with those for comparable terrestrial rocks and meteorites. The possible choice is a wide one. For the purpose of this discussion we have selected those used in Table 7-1: the diabase W-1 (a terrestrial basaltic rock); the average figures for the eucrites (a class of achondrite meteorites that is the nearest analog to the lunar rocks in chemical and mineralogical composition); and the elemental abundances in the carbonaceous chondrites, the least-differentiated class of meteorites, whose composition is generally accepted as the nearest approximation to that of the average nonvolatile matter of the solar system.

In 1923 V. M. Goldschmidt proposed a geochemical classification of the elements into siderophile, chalcophile, lithophile, and atmophile, according to their affinity for metallic iron, for sulfides, for silicates and other oxidic minerals, and for the atmosphere, respectively. He remarked that meteorites, with their metal, troilite, and silicate phases, provide a ready-made "fossilized" distribution experiment. At that time the data on the distribution of the elements in meteorites were few in number and poor in quality. However, the utility of Goldschmidt's classification has been generally accepted, and the data have been vastly improved. Table 7-3 gives the geochemical classification of the elements based on their distribution in the common chondritic meteorites. A few elements appear in more than one group: iron, in particular, is present in the chondrites as metal, in troilite, and in silicates. The presence of troilite and free iron in the lunar rocks indicates that similar geochemical separations occurred during

Table 7-3. Geochemical Classification of the Elements, Based on Their Distribution in the Common Chondrites; Parentheses Indicate a Minor Affinity

Siderophile	Chalcophile	Lithophile	Atmophile
Fe Co Ni	S Se Te	Li Na K Rb Cs	He Ne Ar Kr Xe
Ru Rh Pd	Fe Ag Cd	Be Mg Ca Sr Ba	H N C
Os Ir Pt	Hg Tl Pb	B Al Sc Y La-Lu	
Cu Au Mo	Bi In (Mo)	Si Ti Zr Hf Th	
W Re Ge		P V Nb Ta Mn Fe	
As Sb Sn		O Cr U Zn Ga	
(Ga) (Bi)		F Cl Br I	

their crystallization as in the common chondrites. Hence the information in Table 7-3 is useful in predicting the phase or phases in which a specific element is likely to be concentrated.

Within the large group of lithophile elements, an important controlling factor for the distribution of a specific element is its ionic radius. Elements of similar ionic size tend to replace each other readily in a specific mineral. For example, the ionic radii of barium and rubidium are similar to that of potassium, and these elements concentrate in potash feldspar; the rare earths (and yttrium) have radii similar to that of calcium and are concentrated in calcium-rich minerals, especially apatite; chromium, titanium, vanadium, and iron have similar ionic radii and are combined in a number of complex iron-titanium oxides. Further examples will be discussed under the individual elements.

Hydrogen

One of the most eagerly anticipated geochemical questions to be answered by the Apollo 11 mission was the possible presence of hydrogen in combination in the lunar rocks. Terrestrial igneous rocks always contain measurable amounts in the form of hydroxyl groups in minerals such as amphiboles and micas, as combined or adsorbed water molecules, or possibly in more occult forms. If the lunar rocks were similar, this would not only be a scientific discovery of the first order, but would also have important practical bearing on the possibility of providing future astronauts with some of their requirements of that essential fluid, water.

The lunar rocks have proved to be extremely "dry." Microscopic aqueous inclusions, common in terrestrial igneous rocks, are completely absent. Friedman et al. (1970) report that the water content of the breccias is 150–455 ppm and the fines 810 ppm. Despite all precautions the possibility of contamination with terrestrial water vapor is always present. However, the investigators do not believe that such contamination was serious, since (1) the water is not evolved below 300°C (this would tend to rule out adsorbed water); (2) all the samples, both those exposed to the terrestrial environment and those sealed in vacuum, reacted similarly when heated and gave about the same amounts of hydrogen and water with about the same deuterium content; and (3) the deuterium content of the water is far lower than that to be expected from water picked up from the terrestrial atmosphere—the lunar water contains about 50 ppm deuterium, whereas terrestrial water contains 150–200 ppm.

Beside the water the lunar samples also evolved hydrogen gas in amounts of 40–53 ppm, with a slightly lower deuterium content than the lunar water. The hydrogen gas emitted by the samples may have been present as trapped molecular hydrogen or, alternatively, it may have been generated during heating by the reaction of water with carbon, metallic iron, or ferrous iron present in the silicates. However, all the hydrogen was liberated below 550°C, whereas appreciable amounts of water were released above that temperature, which tends to rule out chemical reactions as a source of the hydrogen.

Helium, Neon, Argon, Krypton, and Xenon

The LSPET team (1969), in its preliminary examination of the Apollo 11 samples, reported that the lunar fines contained vast quantities of the inert gases. This has been confirmed by later investigators; the comparative amounts of these gases in the crystalline rocks and the fines are shown in Table 7-4 (data from Funkhouser et al., 1970).

Possible sources for the rare gases are: (1) inherent, incorporated in the rock at the time of original crystallization; (2) radiogenic, produced by the disintegration of radioactive elements such as uranium, thorium, and ^{40}K; (3) cosmogenic (also known as spallogenic), produced by spallation of other elements through cosmic-ray bombardment: (4) the solar wind; and (5) the addition of gas-rich cosmic dust to the lunar surface. The amounts of the rare gases may be subject to modification by diffusion loss from the Moon; such loss would be most effective for the lighter rare gases, helium and neon.

The evidence indicates that most of these gases in the lunar fines and breccias were derived from the solar wind, while the gases in the crystalline rocks are principally radiogenic and cosmogenic. Hintenberger et al. (1970) have shown that in the crystalline rocks much of the ^4He and essentially all of the ^{40}Ar are radiogenic. Marti et al. (1970) comment that no inherent gases were found in the crystalline rocks, indicating that they were completely outgassed at the time of their formation. The contribution of an input of gas-rich cosmic dust to the lunar fines and breccias is probably very small, at least in comparison to the solar wind component.

Lithium

A considerable amount of information is available on the abundance of this element in lunar samples. The range 9–23 ppm in Table 7-1 excludes a few more extreme figures, which generally are in disagreement with other analyses for the same samples. No consistent difference is apparent between the type A and type B rocks; lithium abundances in the breccias and soil

Table 7-4. Rare Gas Content of Crystalline Rock (10058) and Fines (10010). (Concentrations in units of 10^{-8} cm^3/g, at standard temperature and pressure)

Rock Number	^{3}He	^{4}He	^{20}Ne	^{21}Ne	^{22}Ne	^{36}Ar	^{38}Ar	^{40}Ar	^{84}Kr	^{132}Xe
10	7500	19,000,000	313,000	800	24,000	34,000	6600	41,000	20	4.1
58	58	21,000	64	7.1	12	15	12	3700	0.015	0.007

are generally at the lower end of the range. The lithium content of the lunar materials is comparable with that in the eucrites and in W-1; it is an order of magnitude higher than in carbonaceous chondrites.

No lithium determinations have been made in separated lunar minerals. From crystal chemistry one might predict that the lithium proxies for magnesium and that this element will be present for the most part in pyroxene and olivine.

Beryllium

The data for this element are rather sparse and have been obtained by spark-source mass spectrometry (Morrison et al., 1970) and emission spectrography (Annell and Helz, 1970). The spread in the results is not great, and there is no consistent difference between different rock types. The mean figure for the lunar materials (2 ppm) is comparable with that for W-1 and is at least an order of magnitude greater than for the eucrites and the carbonaceous chondrites (few data are available for Be in meteorites; the figures given in Table 7-1 are not well established). No information is available on beryllium in lunar minerals, but this element probably substitutes for silicon, and aluminum in four-coordination, in which case it is dispersed in pyroxene and plagioclase.

Boron

The limited data for this element have been obtained by neutron activation analysis (Wänke et al., 1970) and spark-source mass spectrometry (Morrison et al., 1970). The range is not great; the mean (2 ppm) is comparable with the meteorite values in Table 7-1 and considerably below that for W-1. In meteorites plagioclase and olivine are the principal carriers of boron (Mason and Graham, 1970), and this is probably the case in lunar materials.

Carbon

In view of the extreme interest in the possible presence of organic matter on the Moon, carbon analyses of the lunar samples are of major importance. The basic data for carbon (and nitrogen) have been provided by Moore et al. (1970), using combustion gas chromatographic techniques by which all the carbon in the samples is converted to CO_2 and determined in this form. Different samples of the same specimen, especially of the crystalline rocks, sometimes gave rather variable results, suggesting some localization of the carbon in specific and irregularly distributed phases (possibly cohenite and/or other iron carbides, which have been identified in the lunar rocks).

The weighted means for individual specimens from Apollo 11 range from 64–230 ppm, with an overall average of 140 ppm. The crystalline rocks average less than 100 ppm, and the breccias and the fines contain significantly greater amounts. Similar results have been obtained on the Apollo 12 materials (LSPET, 1970). Moore et al. point out that the carbon (and nitrogen) in the lunar materials may come from the following sources: (1) indigenous, (2) meteoritic or cometary infall, (3) the solar wind, and (4) contamination. Contamination has, of course, been carefully guarded against and does not appear to be significant in the analyzed samples. The total carbon trends indicate that the values of 64 and 70 ppm found for the lunar basalts are primarily due to indigenous carbon, and the higher contents in the breccias and fines to the carbon introduced by meteorites, comets, and the solar wind. Moore et al. sieved a sample of the lunar fines and found an increase in carbon in the finer sieve fractions as follows: 60–140 mesh, 115 ppm; 140–300 mesh, 210 ppm; and −300 mesh, 500 ppm.

Carbon contents are not available for W-1 or for the eucrites but are certainly very low. The high content of the type I carbonaceous chondrites, averaging 3.5%, suggests that less than 1% of such material added to the lunar surface would account for the increment in the breccias and fines (unless a considerable amount of the added carbon escaped as volatile compounds).

The isotopic composition of lunar carbon has been measured by several investigators. Kaplan and Smith (1970) found consistent differences in δ ^{13}C values in different rock types, as follows: crystalline rocks, −18.8 to −29.8; breccias, +1.6 to +9.2; fines, +17.2 to +20.2. The lunar fines are thus enriched in ^{13}C, relative to the lunar crystalline rocks (and to terrestrial carbon, with a normal range of +2 to −30). Kaplan and Smith speculate that ^{12}C has been preferentially lost from the breccias and fines, possibly by the escape of ^{12}C methane formed by protons in the solar wind reacting with the carbon.

An extensive search for organic compounds in the lunar samples by many investigators employing highly sophisticated techniques has produced negative results. The minute amounts (of the order of 1 ppm) reported can plausibly be ascribed to contamination, either with rocket exhaust products or by terrestrial material.

Nitrogen

The concentration of nitrogen in lunar samples is similar to that of carbon and shows the same trends (Moore et al., 1970). The lowest concentration, 30 ppm, was found in a medium-grained crystalline rock; a fine-grained crystalline rock had 115 ppm, a breccia 125 ppm, and different samples of fines 102 and 153 ppm. As with carbon, the nitrogen in the lunar samples appears to represent a mixture of indigenous material together with meteoritic and solar wind components.

Hintenberger et al. (1970) treated a sample of lunar breccia with dilute H_2SO_4 and found that less than 10% of the total nitrogen was evolved as gaseous N_2, whereas more than 65% was leached out as NH_4^+, possibly indicating the presence of a nitride or nitrides; an additional 15% was present as a nitrate. The isotopic composition of the lunar nitrogen agreed with that of terrestrial nitrogen.

Oxygen

Ehmann and Morgan (1970) have made an extensive series of measurements of oxygen in lunar materials by means of neutron activation analysis. They find discrete ranges and means for each of the major rock types, as follows (wt %): fine-grained crystalline rocks, 37.6–39.3, 38.5; medium-grained crystalline rocks, 38.1–40.5, 39.4; breccias, 39.8–43.4, 41.1; fines, 39.8–42.2, 40.8. The oxygen content is essentially determined by the proportions of the three major minerals: pyroxene, plagioclase, and ilmenite. Since pyroxene and plagioclase have approximately the same oxygen contents (about 44% and 46%, respectively) the different oxygen contents in the lunar rocks reflects differences in their ilmenite content (ilmenite has 31.6% oxygen). The fine-grained crystalline rocks have more ilmenite than the medium-grained rocks, and the breccias and fines have less ilmenite than the crystalline rocks. The lower oxygen content of the lunar materials compared to W-1, the eucrites, and carbonaceous chondrites is also due to their high ilmenite content.

The isotopic composition of the oxygen in the lunar materials has been studied by several investigators. The $^{18}O/^{16}O$ values range from $+5.6$ to $+6.2$ per mil, close to the best average value of $+5.9$ for terrestrial basalts (Fig. 7-1). Onuma et al. (1970) determined $^{18}O/^{16}O$ values for separated plagioclase, pyroxene, and ilmenite; they found that the isotopic distribution signifies equilibrium at 1120°C, which corresponds to the solidus temperature of the lunar basalt. They note that the isotopic composi-

128   THE LUNAR ROCKS

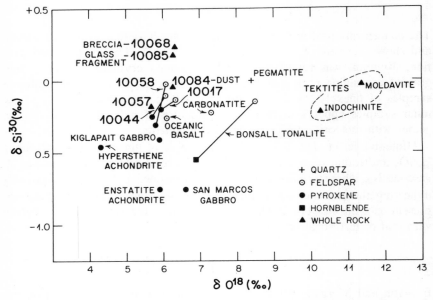

Fig. 7-1. Graph of $\delta\ ^{30}Si$ versus $\delta\ ^{18}O$ for lunar rocks and minerals, with comparative analyses for some meteorites, tektites, and terrestrial igneous rocks (Epstein and Taylor, 1970).

tion for the lunar pyroxene is somewhat higher than for pyroxene from the eucrites, which might indicate a nonlunar origin for these meteorites.

Fluorine

A limited amount of data is available for this element, obtained by spark-source mass spectrometry (Morrison et al., 1970) and by photon activation analysis (Reed et al., 1970). Since the range of values is considerable and the agreement between the two methods, when samples of the same material were analyzed, is not good, the data must be interpreted cautiously. No consistent differences between different rock types is evident. The concentration of fluorine in the lunar materials is comparable with that in W-1 and stony meteorites.

The only fluorine-bearing mineral identified in the lunar materials is apatite, $Ca_5(PO_4)_3(F, Cl)$. Pure fluorapatite contains 3.8% fluorine, but lunar apatite contains less than that because of partial substitution by chlorine. The phosphorus content of the lunar samples could account for up to about 150 ppm fluorine in the form of apatite. Contents of more than 150 ppm, if verified, suggest fluorine in other combinations, possibly substituting for oxygen.

Sodium

A large amount of data on this element is available, since it has been determined in the lunar samples both by standard chemical analyses and by instrumental methods such as atomic absorption and neutron activation analysis. The results are highly consistent, and the range is remarkably narrow, 2600–4000 ppm. Essentially all the sodium in the lunar material is present as plagioclase (or, in the breccias and fines, as glass formed from plagioclase). The sodium contents of the different types of lunar material show no consistent differences, which at first glance is surprising, since the medium-grained rocks contain more plagioclase than the fine-grained rocks, and the breccias and fines more plagioclase (and plagioclase-derived glass) than the crystalline rocks. The answer to this apparent paradox probably lies in the composition of the plagioclase; as a solid solution of $NaAlSi_3O_8$ and $CaAl_2Si_2O_8$ it can vary in sodium content. There also may be consistent differences in the average sodium content in plagioclase from different rock types; specifically, the plagioclase in the fine-grained crystalline rocks may be somewhat more sodic than that in the medium-grained rocks.

Goles et al. (1970) have made the pertinent suggestion that the uniformity of sodium content in the lunar materials reflects some fundamental process, related perhaps to distribution coefficients for sodium between solid and fluid phases. Of possible significance in this connection is a similar uniformity of sodium content in the eucrites, a content which is essentially identical with that of the lunar samples.

Magnesium

Magnesium, being an important component in all the lunar materials, has been determined in a large number of analyses. The range in the Apollo 11 samples, 3.4–5.1%, is quite small, because pyroxene, the principal magnesium-bearing phase, is present in rather uniform amount in all of them. The amount of magnesium in the Apollo 11 rocks is similar to that in W-1 and the eucrites, but only about half that in the carbonaceous chondrites, indicating that the Apollo 11 rocks are relatively depleted in this element compared to undifferentiated meteoritic matter. Magnesium averages about 50% higher in the Apollo 12 rocks, largely because of their much greater olivine content.

Aluminum

Aluminum is an important component in the lunar materials and has been determined in all the complete chemical analyses. In addition, Ehmann and Morgan (1970) have provided a considerable number of determinations by neutron activation analysis for the Apollo 11 samples. Their results show a significant quantification of aluminum content by rock type, as follows (range and mean, wt %): fine-grained crystalline rocks, 3.7–4.2, 4.0; medium-grained rocks, 4.1–5.5, 5.0; breccias, 5.7–7.4, 6.6; and fines, 7.1–7.3, 7.2. Thus the fine-grained rocks contain less aluminum than the medium-grained rocks, and these rocks contain less than the breccias and the fines. Since most of the aluminum is combined in plagioclase, the aluminum content reflects the amount of plagioclase or plagioclase-derived glass. The breccias and fines evidently contain a plagioclase increment additional to that derived from the local crystalline rocks; this has been plausibly ascribed to the infall of plagioclase-rich rocks derived from the lunar highlands. The average aluminum content for the Apollo 12 rocks is notably higher, reflecting their generally higher plagioclase content.

Silicon

As a major component in the lunar samples this element has been determined in all the complete chemical analyses, and a series of neutron activation analyses for the Apollo 11 rocks has been provided by Ehmann and Morgan (1970). Their results for the different rock types are as follows (range and mean, wt %): fine-grained crystalline rocks, 18.3–19.4, 18.9; medium-grained rocks, 17.9–20.1, 18.7; breccias, 18.1–20.6, 19.7; fines, 20.0–20.4, 20.2. Thus the lunar fines and breccias show a slight but consistent enhancement in this element relative to the crystalline rocks, similar to that shown for aluminum, and evidently for the same reason, an increment of plagioclase in the breccias and fines.

The silicon content of the lunar materials is somewhat lower than that of W-1 and the eucrites because the latter contain very little ilmenite or other nonsilicates. The lunar materials contain approximately twice as much silicon as the carbonaceous chondrites, further evidence of their differentiated nature.

Phosphorus

This element is present in all the lunar materials in small amounts: the range is quite limited, about 200–900 ppm, equivalent to 0.04–0.20% P_2O_5. No significant differences have been detected between the different rock types. The principal phosphorus-bearing phase is apatite, although whitlockite has also been recorded, and traces of the phosphide schreibersite in a few samples. The mean content in the lunar material is similar to that in W-1 and the eucrites but is considerably lower than in the carbonaceous chondrites.

Sulfur

Like phosphorus, sulfur is present in all the lunar materials in small amounts. The range is limited, 1200–2400 ppm, and the amounts in the different rock types are not markedly quantized, although the breccias and fines may contain somewhat less than the crystalline rocks. The sulfur is present as troilite, FeS, and the amount corresponds to about 0.3–0.7% of this mineral. The sulfur content in the lunar materials is notably higher than in W-1 and the eucrites but very much lower than in the carbonaceous chondrites.

Kaplan and Smith (1970) report that sulfur from the crystalline rocks has similar isotopic composition to meteoritic sulfur, whereas sulfur from lunar fines is enriched in ^{34}S by 5 per mil or more. They suggest that reaction with protons from the solar wind may favor the conversion of ^{32}S to hydrogen sulfide, which would then be lost from the moon's surface.

Chlorine

This element has been determined in lunar samples by neutron activation analysis (Reed et al., 1970; Wänke et al., 1970) and by spark-source mass spectrometry (Morrison et al., 1970). The neutron activation results are reasonably consistent, whereas those from spark-source mass spectrometry are much higher, up to a factor of 10. We have therefore used only the neutron activation data in Table 7-1. On the basis of these results, chlorine is about ten times less abundant than fluorine in the lunar materials. Some of it is present as apatite, $Ca_5(PO_4)_3(F, Cl)$, but Reed et al. found that in some samples a considerable amount was water-soluble and thus not derived from apatite.

Chlorine appears to be much lower in lunar materials than in terrestrial basaltic rocks such as W-1 or in carbonaceous chondrites; comparable abundances have been found for the eucrites.

Potassium

A large number of determinations is available, since the potassium content of the lunar materials is required to evaluate the contribution of

^{40}K to the heat balance and the production of ^{40}Ar. The data from different investigators are highly consistent, showing a range of 400–2800 ppm and a mean of about 1400 ppm. More significant than the overall mean, however, are the means for the specific rock types of the Apollo 11 rocks. For the fine-grained crystalline rocks (A) the mean is about 2300 ppm, and for the medium-grained rocks (B) about 600 ppm (Table 7-2); for the breccias (C) and fines (D) the mean is about 1200 ppm, an intermediate figure consistent with their being an average mixture of the crystalline rock types.

The contrast between the quantization of potassium contents as against the uniformity of sodium in the crystalline rocks illustrates the significant difference in crystal chemistry between these two otherwise closely related elements. The ionic radius of sodium (0.97 Å) is close to that of calcium (0.99 Å), and hence sodium proxies for calcium in plagioclase. Potassium, a much larger ion (1.33 Å), is not readily accepted into the plagioclase structure and, during magmatic crystallization, accumulates in the residual liquid until its concentration is high enough for the formation of potash feldspar (or it solidifies as a potash-rich glass). The striking difference between type A and B rocks in their potassium content can perhaps be ascribed to the retention of the residual liquid in type A rocks and its expulsion from type B rocks. The sharpness of the quantization is surprising, however; one might expect to find a few crystalline rocks with intermediate potassium values but these appear not to occur—all type A rocks have 2000 ppm or more, whereas the type B rocks all have less than 1000 ppm.

The lunar rocks are notably poorer in potassium than W-1, but have about the same amounts as in oceanic ridge basalts. The eucrites are comparable to the type B rocks, suggesting that they too may have been depleted in potassium by the loss of a residual liquid.

Calcium

A large amount of analytical data is available for this element, since it is a major component in all the lunar samples. The results for the Apollo 11 rocks show a narrow range of calcium content, 7.2–9.0%, equivalent to 10.1–12.6% CaO. Calcium is a major element in pyroxene and plagioclase, two of the three principal minerals in the lunar materials. The amount of pyroxene is fairly constant in all the Apollo 11 material, whereas the amount of plagioclase varies somewhat; however, this variation is only weakly reflected in the calcium content but is much more pronounced in the aluminum content.

The mean calcium content in the lunar materials is very similar to that in W-1 and the eucrites, and much higher than that in the chondrites.

Scandium

Several investigators have measured this element in a considerable number of lunar samples, and their data are in excellent agreement. The range is small, 60–100 ppm; the mean (75 ppm) is about twice as high as the figure for W-1 and the eucrites, and an order of magnitude higher than for the carbonaceous chondrites. Scandium thus shows a marked enrichment in the lunar materials. The crystalline rocks consistently show somewhat higher contents than the breccias and fines. Goles et al. (1970) have found that scandium is concentrated in pyroxene and is strongly depleted in plagioclase; the breccias and fines contain somewhat more plagioclase than the crystalline rocks, and this appears to be responsible for their lower scandium contents.

Titanium

Probably the most remarkable geochemical feature of the Apollo materials is their high titanium content. This totally unexpected phenomenon was predicted by Turkevich et al. (1969) from the results of their alpha-scattering experiment on the Surveyor V landing in Mare Tranquillitatis and confirmed by the initial analyses of the LSPET team (1969). All the Apollo 11 samples (except rare plagioclase-rich fragments in the breccias and fines) have notably high titanium contents, ranging from 4.3–7.4%, with a mean of 5.9%; this is an order of magnitude higher than in W-1 and the eucrites, and two orders of magnitude higher than in the chondrites. Most of the titanium in the lunar materials is present as ilmenite but a small and variable amount is contained in pyroxene. The crystalline rocks contain more ilmenite than the breccias and fines, and hence have higher titanium contents; among the crystalline rocks the fine-grained (type A) contain more ilmenite than the medium-grained (type B).

Preliminary investigation of the Apollo 12 rocks has shown that they are consistently lower in titanium than the Apollo 11 materials, averaging about half as much. They are, nevertheless, still higher than most terrestrial rocks and meteorites.

Vanadium

Several investigators have measured this element in a considerable number of samples, and their results are reasonably consistent, showing a range from about 20–100 ppm, with a mean of 50 ppm. These figures are similar to those for eucrites and chondritic meteorites but considerably lower than for W-1. Most of the vanadium is probably contained in ilmenite, so that one might expect a quantization of vanadium content by rock type, as observed for titanium; however, this is not obvious from the available data, perhaps because of a considerable degree of experimental error in the individual determinations.

Chromium

Many analyses for chromium have been made on the lunar materials, both by classical chemical techniques and by instrumental methods. The results are quite consistent and show a range of 1300–2800 ppm and a mean of 2100 ppm. The fine-grained crystalline rocks generally have high chromium contents and the medium-grained rocks low, with the breccias and fines at an intermediate level. Most of the chromium is present in pyroxene and ilmenite, but trace amounts of chromium-rich spinels have also been identified in some rocks.

The chromium content in the lunar materials is comparable with that in the eucrites and chondritic meteorites but almost 20 times that in W-1.

Manganese

Manganese abundances are remarkably similar to those of chromium, the range being somewhat more restricted, 1500–2400 ppm, and the mean, 1900 ppm, a little lower. Most of the manganese is contained in pyroxene and ilmenite. No consistent difference is evident between the fine-grained and the medium-grained crystalline rocks, but the manganese content of the breccias and fines falls at the lower end of the range, evidently because of their higher plagioclase content.

The manganese content of the lunar materials is comparable with that of W-1 and the chondritic meteorites, and is somewhat lower than that of the eucrites.

Iron

After silicon and oxygen, iron is the most abundant element in the lunar samples. In the Apollo 11 samples, it shows a moderate range, 11.8–15.6%, with a mean of 14.3%. Most of the iron is combined as pyroxene and ilmenite; the principal variable controlling the iron content is the amount of ilmenite. Thus the fine-grained crystalline rocks have higher ilmenite and higher iron contents than the medium-grained rocks; the breccias and fines show lower contents than the crystalline rocks.

The lunar rocks are notably richer in iron than terrestrial basaltic rocks such as W-1, but are similar to the eucrites, and lower in iron than the chondrites.

Cobalt

A large number of cobalt determinations have been published; they show a comparatively limited range, 11–35 ppm, and a mean of 25 ppm. The values for the fine-grained crystalline rocks cluster around 30 ppm, those for the medium-grained rocks around 15 ppm; however, the breccias and fines, instead of falling in the intermediate range, also have about 30 ppm. The cobalt in the crystalline rocks is indigenous to the Moon, whereas the breccias and fines contain a meteoritic increment. If we assume that the breccias and fines contain approximately equal amounts of fine and medium-grained crystalline rocks, these would contribute about 20 ppm, and the additional 10 ppm represents meteoritic input, which is consistent with the input of other trace elements of meteoritic derivation. Most of the cobalt is probably contained in the metal particles in the lunar materials.

Cobalt is somewhat higher in the lunar rocks than in the eucrites but is somewhat lower than in W-1, and about 20 times less than in chondritic meteorites. The Fe/Co ratio is 380 in the carbonaceous chondrites and 5700 in the lunar rocks, indicating a tremendous depletion in this element.

Nickel

This element provides the clearest evidence for a meteoritic increment in the lunar breccias and fines. Analyses of the crystalline rocks report 3–24 ppm in different samples, whereas for the breccias and fines the analyses report 150–280 ppm. The nickel in the breccias and fines is contained in sparsely and irregularly distributed particles of nickel-iron. Sample differences probably account for much of the range shown by these materials. The metal in the crystalline rocks is reported to contain little or no nickel and is present in such small amounts that it may not account for all the nickel in these rocks.

The concentration of nickel in the lunar crystalline rocks is similar to that in the eucrites and is considerably lower than in W-1. It is about three orders of magnitude lower than in the chondritic meteorites, evidence of extreme depletion in this and other siderophile elements. The Fe/Ni ratio in the lunar crystalline rocks is about 12,000, in contrast to about 18 for the carbonaceous chondrites; an extremely effective separation process must have been responsible for this fractionation.

Copper

Data for this element have been published by several groups of investigators and are generally in good agreement, although a few anomalously high values have been omitted in arriving at a range of 4–25 ppm, with a mean of 11 ppm. There is a general trend for the figures for the lunar crystalline rocks to group around the lower end of the range, and those for the breccias and fines to group around the upper end. Meteoritic nickel-iron averages about 170 ppm copper (the amount, however, is quite variable from one meteorite to another), so that a small addition of this material could account for the somewhat enhanced figures for the breccias and fines.

The amount of copper in the lunar materials is similar to that in the eucrites, and about one-tenth that in W-1 and in the chondritic meteorites.

Zinc

This element has been determined by several groups of investigators, using a variety of analytical techniques. Their results are generally in good agreement, although some anomalously high figures have been disregarded in arriving at the range of 2–40 ppm, with a mean of 15 ppm. Keays et al. (1970) ran an extensive series of measurements on the Apollo 11 materials, and found a marked and consistent difference between the crystalline rocks (A and B) and the breccias (C) and fines (D); their averages for these types are: A and B, 1.6 ppm; C, 29 ppm; and D, 21 ppm. They attribute the enhanced zinc content of the breccias and fines to the addition of meteoritic material. This explanation, eminently satisfactory for siderophile elements, is less convincing for zinc, since this element is low in iron meteorites [Smales et al. (1967) report 0.28–42 ppm in 67 irons, 34 having less than 1 ppm], and even the carbonaceous chondrites, averaging 320 ppm, are an inadequate source.

The zinc content in the lunar crystalline rocks is similar to that in the eucrites, and notably lower than the content in W-1 and the carbonaceous chondrites.

Gallium

Numerous figures for the Apollo 11 samples are available from several groups of investigators; their results are in good agreement and show a remarkably small range, 3–6 ppm, with a mean of 4.5. Keays et al. (1970) recognize a difference between the mean for the crystalline rocks (3.68

ppm) and for the breccias (5.82 ppm) and fines (5.30 ppm), but the difference is small and is not evident in the results of other investigators. If there is an enrichment of gallium in the breccias and fines, it may be due to the somewhat higher plagioclase content of these materials, rather than to meteorite infall, as suggested by Keays et al.

The gallium content of the lunar materials is comparable to that of the eucrites, somewhat lower than that in carbonaceous chondrites and in terrestrial basaltic rocks such as W-1. Terrestrial basaltic rocks contain much more plagioclase than the lunar materials; this may be largely responsible for their higher gallium contents.

Germanium

This element has been determined in lunar materials by Wasson and Baedecker (1970), and Wänke et al. (1970), using neutron activation analysis; by Morrison et al. (1970) using spark-source mass spectrometry. Wänke et al. report less than 1 ppm in three crystalline rocks and 1.4 ppm in the fines. Wasson and Baedecker found 0.39 ppm in the fines, 0.24–0.41 ppm in four samples of breccia, and an upper limit of 0.04 ppm in another; and 0.06 ppm in two crystalline rocks and less than 0.07 ppm in a third. The figures of Morrison et al. are much higher and show no consistent pattern.

Since germanium is an extremely siderophile element under the reducing conditions of lunar crystallization, the low content reported by Baedecker and Wasson in the crystalline rocks is consistent with the impoverishment of these rocks in other siderophile elements such as nickel. The higher content in the breccias and fines can plausibly be ascribed to meteoritic increment.

Arsenic

The data on this element are scanty. Morrison et al. (1970) report approximate figures of 0.03–0.09 ppm for eight samples, with no consistent pattern between the different rock types. Smales et al. (1970) report 0.03 ppm in lunar fines and 0.01 ppm in a breccia. Arsenic is a siderophile element under the reducing conditions of lunar crystallization and should show the same impoverishment in the crystalline rocks as does germanium. The presence of 1% meteoritic nickel-iron in lunar breccias and fines would add about 0.1 ppm As to these materials; from the sparse data at present available there is no convincing evidence of this increment but further work is probably needed.

Selenium

Data on this element have been provided by Morrison et al. (1970) and Haskin et al. (1970). Their results indicate about 0.4–0.8 ppm in the crystalline rocks, and about 1 ppm in breccias and fines. The very low selenium content correlates with the low sulfur content in the lunar materials. The mean S/Se ratio is 2100; this ratio is 2300 in the carbonaceous chondrites, which indicates no marked fractionation of selenium from sulfur in the lunar materials.

Bromine

Data for this element have been provided by Keays et al. (1970) and Reed et al. (1970). Bromine abundances are low, of the order of 100 ppb, and appear to vary in different samples of the same specimen; for example, the following figures (ppb) are given for samples of lunar fines: 126, 159, 187 (Keays et al.); 51, 230 (Reed et al.). Keays et al. give a mean of 31 ppb for the crystalline rocks, 94 ppb for the breccias, and 157 ppb for the fines; but in view of the variability within each rock type and the paucity of the data, these differences may be more apparent than real.

Rubidium

There is a large amount of data on this element, and many of the Apollo 11 samples have been analyzed by several groups of investigators, generally with consistent results. The characterization of two distinct groups of crystalline rocks by discrete trace-element levels is particularly well marked for rubidium, as can be seen in Table 7-2; one group, essentially the fine-grained (type A) rocks, has a range of 5.2–6.2 ppm, with a mean of 5.7 ppm; whereas the other group of medium-grained rocks (type B) has a range of 0.5–1.2 ppm, with a mean of 0.8 ppm. The two groups also have distinctive K/Rb ratios, for the first group around 400 and for the second group 700–900. As might be expected, the breccias and fines have rubidium contents of 2–4 ppm, intermediate between these two groups.

Rubidium, with an ionic radius of 1.47 Å, is geochemically coherent with potassium (1.33 Å). The rubidium in the lunar materials is presumably concentrated in potash feldspar and potash-rich glass. The lunar materials are about ten times richer in rubidium than the eucrites and have about the same amount as the carbonaceous chondrites; however, they contain much lower concentrations than terrestrial basaltic rocks such as W-1.

Strontium

A large number of determinations are available for this element and show a rather narrow range, 110–220 ppm, with a mean of 170 ppm. No systematic differences between the different rock types are evident. Philpotts and Schnetzler (1970) have measured this element in concentrates of pyroxene, plagioclase, and opaque minerals from the lunar rock; they find that it is strongly enriched in the plagioclase, as is also the case in terrestrial igneous rocks and in stony meteorites.

The amount of strontium in the lunar materials is similar to that in W-1, about twice that in the eucrites, and 20 times that in the carbonaceous chondrites. The lunar rocks are thus notably enriched in strontium with respect to the stony meteorites.

Yttrium

Compston et al. (1970) and Annell and Helz (1970) have reported extensive series of measurements of yttrium in Apollo 11 materials; their results on the same specimens are in excellent agreement. Compston et al. give an overall range of 73–168 ppm; Annell and Helz, 81–165 ppm. Compston et al. point out that yttrium shows the same quantization between two groups of crystalline rocks as does rubidium; the fine-grained rocks have around 160 ppm, whereas the medium-grained rocks have considerably lower contents (73–134 ppm).

Yttrium is one of the elements, like titanium and zirconium, that is significantly more abundant in the lunar rocks than in terrestrial basalts or in stony meteorites.

Zirconium

The LSPET account (1969) reported notably high zirconium contents in the Apollo 11 materials, which has been confirmed in numerous analyses by later investigators. The figures range from 180–660 ppm, with a mean of 370 ppm. Table 7-2 shows that there is a somewhat greater concentration (375–644 ppm) in the fine-grained crystalline rocks than in the medium-grained rocks (194–380 ppm). Much of the zirconium is evidently in solid solution in ilmenite; Arrhenius et al. (1970) record 300–3000 ppm in this mineral, with large variations from grain to grain and within single grains. The zirconium not removed in ilmenite evidently remained in the lunar magma until crystallization of the residual liquid, when it formed baddeleyite (ZrO_2) and zircon ($ZrSiO_4$).

Zirconium concentrations in the lunar materials are about an order of magnitude higher than in the eucrites, and about four times higher than in terrestrial basaltic rocks such as W-1.

Niobium

Annell and Helz (1970) and Compston et al. (1970) have reported a series of mutually consistent measurements on niobium in lunar materials. Their results give a range of 14–31 ppm and a mean of 21 ppm, with no significant variation between the different rock types. These concentrations are somewhat higher than in terrestrial basaltic rocks such as W-1, and at least an order of magnitude higher than for stony meteorites, indicating that niobium is notably enriched in lunar materials.

Molybdenum

Morrison et al. (1970) have provided data on this element in eight samples of lunar material. In four crystalline rocks they found a uniform content of 0.4 ppm; two breccias and a sample of fines gave 0.7 ppm, and another breccia 0.4 ppm. In stony meteorites molybdenum is siderophile with minor chalcophile affinity. The increased amount in lunar breccias and fines compared to the crystalline rocks is consistent with the introduction of a minor amount of meteoritic material.

The molybdenum content of the lunar rocks is similar to that of W-1, and somewhat lower than that of the carbonaceous chondrites.

Ruthenium, Rhodium, and Palladium

For this triplet of closely related siderophile elements, data are available only for palladium. Keays et al. (1970) have provided analyses in eight samples, covering the different types of lunar materials. Their averages for the different types, in ppb, are: crystalline rocks, 3.4; breccias, 9.9; and fines, 9.1. This pattern is consistent with that for other siderophile trace elements, showing an enrichment in breccias and fines by the addition of meteoritic matter.

On the basis of the relative abundances of these three elements, it can be predicted that ruthenium will be present in lunar rocks at about the same concentration as palladium, and rhodium at about one-third of this. Compared to carbonaceous chondrites, palladium is depleted by at least two orders of magnitude in the lunar rocks.

Silver

This element is present at very low concentrations in the lunar samples, as indicated by the analyses of Keays et al. (1970). They reject three of their determinations as anomalously high, which they ascribe to contamination from the silver-indium vacuum seals on the sample return containers. Apart from these determinations, their results give the following averages for the different rock types, in ppb: crystalline rocks, 1.46; breccias, 23.6; and fines, 8.72; the breccia figure is a single determination and may not be representative. Keays et al. believe that the enhanced concentrations in breccias and fines are due to the addition of material of carbonaceous chondrite composition; if this is so, the silver content of the lunar fines could result from the input of about 2% of such material.

Cadmium

The data provided for this element by Keays et al. (1970) are similar to their results for silver, except that cadmium concentrations are about four times greater. They find the same marked increase in the lunar breccias and fines over the crystalline rocks, and ascribe it to the same cause, that is, the addition of material of carbonaceous chondrite composition. Their average figures for the different materials are, in ppb: crystalline rocks, 4.64; breccias, 92; and fines, 39.4.

Indium

A number of investigators have measured indium in the lunar samples, but many of the results are suspect because of the possibility of contamination from the silver-indium vacuum seals on the sample return containers; Keays et al. (1970) consider all their results questionable on this account. However, a review of the available data indicates a similar pattern for indium as for the neighboring elements silver and cadmium: very low concentrations, about 3 ppb, in the crystalline rocks, and considerably higher figures, about 50 ppb, for the breccias and fines. All three elements, silver, cadmium, and indium, have concentrations in the lunar rocks about two orders of magnitude lower than in the carbonaceous chondrites, and are also severely depleted in comparison to terrestrial basaltic rocks such as W-1.

Tin

Morrison et al. (1970) have determined this element in a few lunar samples by spark-source mass spectrometry. In three samples of crystalline rocks they found 0.4, 0.6, and 1.2 ppm; in one breccia 0.3 ppm, and in a sample of fines 0.7 ppm. Tin is a siderophile element in meteorites, so that one would expect to find enhanced concentrations in the lunar breccias and fines, as shown by other siderophile elements. This is not apparent from the above data, but additional analyses are clearly desirable.

The tin content of the lunar rocks appears to be somewhat less than in terrestrial basaltic rocks such as W-1, and the carbonaceous chondrites.

Antimony

This element has been determined in eight lunar samples by Morrison et al. (1970), using neutron activation analysis. Three samples are reported as containing 0.01 ppm, five as containing 0.005 ppm. No significant differences are apparent between the different rock types. Antimony is a siderophile element under reducing conditions, and its very low abundance in the lunar rocks is consistent with their general depletion in these elements.

Tellurium

Ganapathy et al. (1970) have made a few determinations on Apollo 11 rocks by neutron activation analysis. Four specimens of crystalline rocks show low and uniform contents (8–13 ppb), whereas two breccia samples are considerably higher in this element (72 and 73 ppb), further evidence for a meteoritic increment in the latter rocks.

Iodine

This element has been determined in six lunar samples by Reed et al. (1970), using neutron activation analysis. Their figures cover a wide range, from 0.006–1.45 ppm, and are extremely variable even for a specific rock type. No meaningful average can be derived from these results, and the phase or phases within which the iodine resides is unknown.

Cesium

Two extensive series of measurements have been provided by Keays et al. (1970), and by Gast and Hubbard (1970). Their results for different rock types are extremely consistent and show the same clear division of the crystalline rocks into two groups, as was evidenced by the data for potassium and rubidium. For the different rock types the averages are (in ppm): fine-grained crystalline, 0.16; medium-grained crystalline, 0.03; breccias, 0.13; and fines, 0.10. The fine-grained crystalline rocks contain about ten times more cesium than the eucrites, but only about one-sixth as much as terrestrial basaltic rocks like W-1.

Barium

A large amount of data is available for this element, provided by several teams of investigators using a variety of analytical techniques; where different samples of the same specimen were analyzed, the results are usually in good agreement. The overall range (omitting some extreme figures) is 70–340 ppm, with a mean of 200 ppm. However, as Table 7-2 shows, barium is an element that is strongly fractionated between the two groups of crystalline rocks: the fine-grained (type A) average 300 ppm, the medium-grained (type B) 95 ppm. The breccias and fines are intermediate, averaging about 180 ppm.

The LSPET team (1969) reported that barium was considerably enriched in the Apollo 11 samples in comparison to the chondritic abundance; it is also enriched by a factor of six over the abundance in the eucrites. The concentration in the lunar material is similar to that in W-1.

Rare-Earth Elements

The rare-earth elements (REE) are a geochemically coherent group, and studies of their relative abundances in terrestrial rocks have contributed greatly to the understanding of igneous (and sedimentary) processes. The data on yttrium and ytterbium provided by the LSPET team (1969) indicated the probability that the Apollo 11 rocks were notably enriched in the rare-earth elements, and this has been confirmed by numerous analyses. These elements show the same quantization in the crystalline rocks as for other elements such as potassium, rubidium, zirconium, and barium (Table 7-2). Taking cerium as an example, the fine-grained crystalline rocks (type A) show a range of 75–108 ppm, with a mean of 83 ppm; whereas for the medium-grained rocks (type B) the range is 23–48 ppm and the mean 38 ppm. The breccias and fines have REE contents intermediate between those of the two groups of crystalline rocks. As in the stony meteorites, these elements are probably concentrated in the calcium phosphate minerals apatite and whitlockite, with lesser amounts in the pyroxenes.

The individual rare-earth elements show a uniform degree of enrichment in the lunar materials, except for europium. This is dramatically

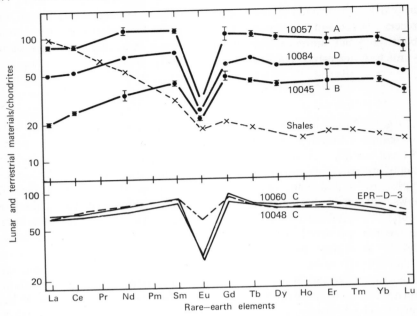

Fig. 7-2. Comparison diagram for rare-earth abundances for Apollo 11 materials (solid lines), a composite of North American shales (upper dashed line), and a submarine andesite (EPR-D-3) from the East Pacific Rise (lower dashed line). (Haskin et al., 1970.)

shown in Fig. 7-2. This diagram (Haskin et al. 1970) compares REE abundances for some lunar samples with those for a composite of nine chondritic meteorites, the abundances for the individual elements in the chondrite composite being taken as unity. The curves for rocks 10057 and 10045 represent the extremes of REE concentrations among the nine samples analyzed; these lunar rocks are thus enriched in the rare-earth elements by factors of 20–100 over the chondrite composite. The curves are quite smooth except at europium, which is strongly depleted in comparison to the neighboring elements (although still greatly enriched over the amount in the chondrite composite). This depletion is evidently linked with a unique geochemical feature of europium: under highly reducing conditions (such as were present during the crystallization of the lunar rocks), this element exists as the Eu^2 ion. This ion is considerably larger than the Eu^3 ion (1.15 Å as against 0.98 Å), which affects its geochemical behavior; unlike the other rare-earth elements, it preferentially enters the plagioclase structure. Philpotts and Schnetzler (1970) have demonstrated this for separated minerals from the lunar rocks.

The marked depletion of europium shown by the Apollo 11 rocks suggests that, unless the original material accreting to form the Moon was similarly depleted in this element (which seems unlikely), there should be rock types elsewhere on the Moon showing a corresponding enrichment. The most likely types would be plagioclase-rich rocks. It is thus highly significant that fragments of such rocks were found in the Apollo 11 breccias and fines, leading, along with other evidence, to the suggestion that the lunar highlands are largely composed of such material.

Hafnium

The data for hafnium show a moderate range of concentration, 7–18 ppm, with a mean of 13 ppm. As might be expected, hafnium shows the same tendency as zirconium to be more concentrated in the fine-grained crystalline rocks, but the effect is not so pronounced; the fine-grained rocks average near the top of the range at about 17 ppm, whereas for the medium-grained rocks the average is about 10 ppm. The overall Zr/Hf ratio (by weight) in the lunar samples is 26, close to the ratio of 28 in the carbonaceous chondrites. The absolute abundance of hafnium in the lunar materials is greater by at least an order of magnitude than in the stony meteorites.

Tantalum

The reported concentrations in the lunar materials range from 1.0–2.7 ppm, with a mean of 1.7 ppm; there are no consistent differences between the different rock types but the data are rather sparse for such a comparison. Like hafnium, tantalum shows enhanced concentrations in lunar materials by at least an order of magnitude over those in stony meteorites.

Tungsten

Tungsten abundances in ten samples of lunar materials have been provided by Wänke et al. (1970) and Morrison et al. (1970), using neutron activation analysis. Their results are in good agreement and give a range of 0.13–0.43 ppm, with a mean of 0.31 ppm. Tungsten is a strongly siderophile element in meteorites and might be expected to show an enrichment in lunar breccias and fines from the meteoritic input. This is not evident from the data available, there being no consistent difference between the different rock types. However, it appears that the concentration of tungsten in the lunar crystalline rocks may be similar to that in meteorites, in which case the addition of meteoritic material would not lead to enhancement of this element.

Rhenium

Herr et al. (1970) have reported 0.011 ppm of this element in a sample of the lunar fines. Since rhenium is a siderophile element in meteorites, the fines are probably enriched in rhenium by the meteoritic input; this figure may be considerably greater than the concentrations in the lunar crystalline rocks.

Osmium, Iridium, and Platinum

For these closely related siderophile elements, no data are available for platinum, and only a little for osmium. Keays et al. (1970) provide iridium data for eight lunar rocks but eliminate two measurements as being unacceptably high, probably because of laboratory contamination. They report the following averages, in ppb: crystalline rocks, 0.066; breccias, 8.03; and fines, 7.14. These figures show clearly the effect of a meteoritic increment in the lunar breccias and fines. Their results have been confirmed by Wasson and Baedecker (1970), who report 0.01 and 0.46 ppb in two crystalline rocks, a mean of 7.0 ppb in the breccias, and 10.7 ppb in the fines.

For osmium, Lovering and Butterfield (1970) report about 0.3 ppb in the crystalline rocks, and about 8 ppb in the fines.

Gold

Keays et al. (1970) have published analyses for gold on eight lunar rocks. They find the same pattern as for iridium, the concentrations in the breccias and fines being almost two orders of magnitude greater than in the crystalline rocks. They give the following averages, in ppb: crystalline rocks, 0.041; breccias, 3.04; and fines, 2.85. Wasson and Baedecker (1970) report comparable results: 0.16, 0.72, and an upper limit of 0.31 ppb in three crystalline rocks, a mean of 1.8 ppb in the breccias, and 1.4 ppb in the fines.

Mercury

Reed et al. (1970) analyzed six samples of lunar materials and found a range of 0.6–13 ppb, with the maximum in a fine-grained crystalline rock. They comment that these concentrations are significantly lower in general than those in the calcium-rich achondrites (18–9000 ppb) or in chondrites, and approach more nearly the values observed in some terrestrial rocks.

Thallium

Data have been provided for this element by Keays et al. (1970). They find the same distribution pattern between the different rock types as they found for several other elements such as gold and iridium. The following averages were obtained for the different rock types, in ppb: crystalline rocks, 0.63; breccias, 2.76; and fines, 2.22. The higher concentrations in the breccias and fines are plausibly ascribed to a meteoritic increment. Ganapathy et al. (1970) note a correlation between thallium and potassium abundances, and suggest that thallium may be lithophile in the lunar rocks.

Lead

Tatsumoto and Rosholt (1970) and Silver (1970) have provided extensive and consistent data on lead in the lunar materials. The overall range is 0.29–1.79 ppm, with an average of 1.2 ppm. Lead shows the same distribution pattern as potassium and rubidium, being concentrated in the fine-grained crystalline rocks (average 1.6 ppm), compared to the medium-grained rocks (average 0.5 ppm).

The lead in the lunar materials is highly radiogenic. This suggests that practically all the primordial lead was lost (by vaporization at high temperatures) before the Moon aggregated; the lead now present has been formed almost entirely by the breakdown of uranium and thorium. Since uranium and thorium are also enriched in the fine-grained crystalline rocks relative to the medium-grained rocks, this would explain the similar distribution pattern for lead.

Bismuth

Keays et al. (1970) have provided data for this element, whose concentration and pattern of distribution closely resemble that of thallium. They find an almost tenfold increase in concentration in the lunar breccias and fines compared to the crystalline rocks. The averages for the different types are, in ppb: crystalline rocks, 0.33; breccias, 2.20; and fines, 1.63.

Thorium

Special interest attaches to thorium and uranium because of their radioactivity; as a result the analytical data are quite extensive. In the Apollo 11 rocks, both of these elements show the specific distribution pattern characteristic of a number of trace elements (Table 7-2), namely, a consistently higher concentration in one group of crystalline rocks (approximately the type A or fine-grained rocks) than in the other (the type B or medium-grained rocks). Results by different investigators are generally in good agreement and show a range of 3.3–4.1 ppm (mean 3.5 ppm) for the type A rocks, and a range of 0.53–1.1 ppm (mean 0.84 ppm) for the type B rocks. The breccias average 2.6 ppm and the fines 2.2 ppm, intermediate as might be expected to the two groups of crystalline rocks.

Thorium is lower in the Apollo 12 crystalline rocks than in the Apollo 11 samples and ranges from 0.77–1.20 ppm, with an average of 0.91 ppm. It is considerably higher in the fines and microbreccias, ranging from 6.0–13.2 ppm, and averages 9.1 ppm. One high-silica rock (12013) is much higher in thorium than any other Apollo samples and most terrestrial granites; it contains 34.3 ppm.

Experience with stony meteorites indicates that thorium (and uranium) are concentrated in the calcium phosphate minerals apatite and whitlockite. Both minerals have been identified in the Apollo 11 rocks and are probably the seat of much of these elements.

Uranium

The discussion of thorium applies equally to uranium. For the Apollo 11 samples, the quantization in two groups of crystalline rocks is equally marked. For the fine-grained rocks the range is 0.74–0.87 ppm, with a mean of 0.85 ppm; for the medium-grained rocks the range is 0.16–0.33 ppm, with a mean of 0.24 ppm. The Apollo 12 crystalline rocks contain considerably less uranium than the Apollo 11 samples. Preliminary results give a range of 0.21–0.31 ppm, with an average of 0.24 ppm for seven samples. On the other hand, the Apollo 12 breccias and fines are much higher in uranium than any of the Apollo 11 samples, ranging from 1.5–3.4 ppm, with an average of 2.3 ppm. One of the most remarkable Apollo 12 samples (12013), termed a feldspathic differentiate (LSPET, 1970), contains 10.7 ppm, a value higher than for most granites, the terrestrial rock series highest in uranium.

Tatsumoto and Rosholt (1970) have found that the uranium isotopic ratio ($^{238}U/^{235}U$) is the same as for terrestrial rocks, within the experimental error. The Th/U ratios for both the Apollo 11 and Apollo 12 rocks range from about 3.4 to 4.0. Terrestrial Th/U ratios for igneous rocks are not markedly different, and fall mainly between 2.7 and 6.0.

Anderson and Phinney (1967) predicted uranium contents of about 0.6 ppm in lunar surface rocks, which they presumed to be of basaltic composition. This prediction, which is remarkably close to values actually measured in the Apollo 11 and 12 igneous rocks, was based on a number of inferences about the physical and chemical composition of the lunar interior and its probable thermal history.

This element-by-element review of the geochemistry of the Apollo 11 samples develops some significant insights into their origin and development. The minor and trace elements clearly define two discrete groups of crystalline rocks, which closely parallel the type A (fine-grained) and type B (medium-grained) petrographic division established by the LSPET team (1969). The fine-grained rocks show uniformly higher concentrations of potassium, rubidium, cesium, yttrium, zirconium, hafnium, barium, the rare earths, lead, thorium, and uranium. The breccias and fines, which on superficial examination might be considered simply a comminuted and partly melted mixture of the crystalline rocks, clearly show the geochemical effects of meteoritic input, the solar wind, and the addition of materials related to the local crystalline rocks but of distinctive composition. The latter increment is recognizable by higher calcium and aluminum and lower titanium than the average for the crystalline rocks, and shows up in the calculated norms as a higher plagioclase content (35–40%, as against 25–30% in the crystalline rocks). The solar wind component is particularly evident in relatively enormous enrichment of the rare gases in the breccias and fines; most of the hydrogen, a small amount of the carbon, and probably a little nitrogen can also be ascribed to introduction by the solar wind. The meteoritic increment is macroscopically evident as occasional fragments and pellets of nickel-iron in the breccias and fines; it is geochemically evident in their relatively high concentrations of the siderophile elements. Thus nickel is about 20 times more abundant in the breccias and fines than in the crystalline rocks, and enrichment factors of up to 100 have been found for other siderophile elements. Not only the siderophile elements, but also the chalcophile elements are markedly enriched in the breccias and fines. Some part of the siderophile elements has been added to these materials in the form of meteoritic nickel-iron; using a figure of 10% for the average amount of nickel in meteoritic nickel-iron, the addition of 0.2% of this would account for the 200 ppm nickel in the breccias and fines. However, Keays et al. (1970) have shown that the overall pattern of trace element enrichment in the breccias and fines, especially the increment in silver, gold, bismuth, bromine, iridium, palladium, and thallium, is best explained by the addition of about 2% of material with the composition of carbonaceous chondrites (Fig. 5-9). The addition of such material would also introduce correspondingly large amounts of carbon and sulfur, not now present in the breccias and fines; however, these elements could conceivably have been removed as gaseous compounds with hydrogen and oxygen.

Some insight into the chemical fractionations that may have taken place during the evolution of the lunar materials can be obtained by comparison with elemental abundances in the carbonaceous chondrites. The carbonaceous chondrites, specifically the type I class, have elemental abundances that probably reflect the average composition of solar system material for the nonvolatile elements. A comparison between these abundances and those for the Apollo 11 rocks can thus reveal the elemental fractionations and perhaps suggest the processes that produced them. The comparison, for those elements for which adequate data are available, is given in Table 7-5 and illustrated in Fig. 7-3.

The extent of these fractionations is immediately apparent. Titanium is over 100 times more abundant in the Apollo 11 rocks than in the carbonaceous chondrites, whereas gold and iridium are over 1000 times less abundant. Some generalizations can readily be made:

1. The highly enriched elements (factors of 10 or more) are all lithophile.
2. All the chalcophile (except lead) and siderophile elements are highly depleted (factors of 0.05 or less).

Table 7-5. Ratios of Elemental Abundances in Apollo 11 Crystalline Rocks to Corresponding Abundances in Type I Carbonaceous Chondrites

>100:Ti (140).

50–100:La-Lu, except Eu ($\sim$100); Ta (85); Y (75); U (50); Th (50); Ba (50); Be (50).

10–50:Nb (42); Zr (41); Hf (40); Eu (38); Sr (21); Li (11).

2–10:Ca (7.3); Al (6.6); K (2.7).

0.5–2:Si (1.9); Rb (1.5); Mn (1.1); Cr (0.95); V (0.88); O (0.87); Fe (0.78); F (0.73); Na (0.60); Cs (0.52).

0.05–0.5:Mg (0.47); Pb (0.41); B (0.40); Ga (0.38); P (0.36).

0.005–0.05:Cu (0.05); Sb (0.05); Cl (0.05); Co (0.04); In (0.04); N (0.04); S (0.03); As (0.03); Se (0.02); Br (0.006); Pd (0.006); Cd (0.005); Tl (0.005).

<0.005:Ag (0.004); C (0.004); Bi (0.003); Ge (0.003); Ni (0.001); Au (0.0002); Ir (0.0002).

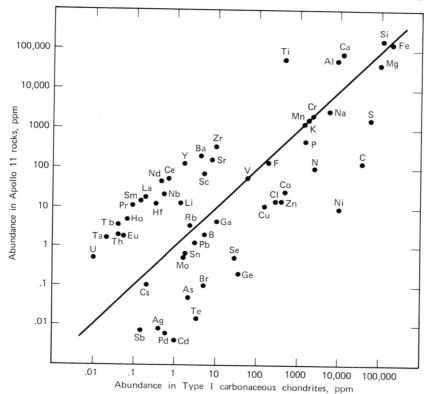

Fig. 7-3. Plot (logarithmic scale) of elemental abundances in Apollo 11 crystalline rocks versus abundances in Type I carbonaceous chondrites. Points falling along the diagonal line represent elements of equal abundance in these materials; points above the line represent elements relatively enriched in the lunar rocks, whereas points below the line represent elements relatively depleted in these rocks. The distance of a point from the diagonal line is a measure of the relative enrichment or depletion. Note the enrichment of refractory lithophile elements and the depletion of siderophile and chalcophile elements in the lunar rocks.

3. Volatile elements (carbon, nitrogen, sulfur, chlorine, selenium, and bromine) are also highly depleted.

Of course, a moot question in this discussion is the extent to which the Apollo 11 rocks are unique to their particular location and to what extent they may reflect the overall geochemistry of lunar materials. The instrumental analyses of three lunar sites (two mare and one highland) by Turkevich and his co-investigators have given very similar results, except for iron and titanium, which are notably less abundant at the highland location. Preliminary investigations on the Apollo 12 collections have shown that they consist of similar rock types to the Apollo 11 materials, the most notable geochemical difference being that titanium is approximately half as abundant; the range of composition is greater in Apollo 12 materials. Exotic fragments in the Apollo 11 breccias and fines are mostly plagioclase-rich rocks, but they are probably genetically related to the basaltic crystalline rocks. On present information, the generalizations set out above may well be valid for most of the lunar surface rocks.

All the evidence indicates that the lunar rocks are products of partial fusion and magmatic crystallization (modified for the breccias and fines by mechanical breakup, shock melting, and the addition of exotic material). There is no indication that they have ever been subjected to weathering and sedimentary processes of the types that have been so pervasive in the geochemical alteration of the crustal rocks of the Earth. The mineralogy of the Apollo 11 basaltic rocks indicates that they crystallized from a lunar magma under very dry, extremely reducing conditions; the partial pressure of oxygen was of the order of 10^{-13} atmospheres, more than five orders of magnitude lower than that for terrestrial basaltic magmas.

The very low content of siderophile elements, even in the rare metal particles in the Apollo 11 crystalline rocks, indicates that the parent magma was probably already depleted in these elements. Either the Moon formed from material already impoverished or these elements were removed during lunar evolution. The obvious possibility is removal by the segregation of a nickel-iron core similar to that of the Earth, carrying with it the siderophile elements and possibly the chalcophile elements also (as included droplets of sulfides). The presence of a core in the Moon appears to be unlikely on geophysical grounds, but the formation of even a small core under highly reducing conditions might provide a very effective mechanism for producing the observed impoverishment of siderophile and chalcophile elements in the surface rocks.

Turning now to the highly enriched elements, we find that not only are they all lithophile, but (except for titanium) they are all elements which do not readily substitute for the major elements in the common minerals: pyroxene, plagioclase, and ilmenite. Their concentration in the original magma may have been quite low, but they largely remained in solution during the crystallization of the common minerals, and thus became enriched in the residual liquid. In other words, the Moon as a whole may not contain unusual amounts of these elements; their concentration in the Apollo 11 rocks suggests that these rocks are end products of an extensive fractionation process.

The abundance pattern of minor and trace elements provides some indications that the material that formed the Moon was depleted in volatile elements prior to aggregation. For example, chlorine and bromine are notably depleted, whereas fluorine is not. Chlorides and bromides of many of the depleted elements are comparatively volatile but if formed on the present Moon would condense on the surface; loss by escape from the Moon would require unreasonably high temperatures, whereas loss in the preaccretion stage would take place readily. Lead is peculiarly significant in this regard. The concentration of this element is much higher than of any other chalcophile element, and its isotopic composition shows that it must be almost completely radiogenic. What happened, then, to the non-radiogenic lead that was presumably present in the original material? A reasonable explanation is that it was largely removed as volatile compounds prior to aggregation of the Moon, and the lead we now observe in the lunar materials has been almost entirely generated within the Moon by the decay of uranium and thorium.

The high titanium content of the Apollo 11 materials remains something of an enigma. Already, from Apollo 12, it appears that this may be a characteristic of mare rocks. It seems peculiar that this titanium enrichment is not accompanied by a parallel enrichment in the closely related elements, vanadium, chromium, and manganese. However, examples are known of extreme titanium enrichment in igneous rocks on Earth (leading to the formation of economically important deposits of ilmenite; such deposits are associated with pyroxene gabbros and anorthosites analogous in composition with some of the lunar crystalline rocks). Olsen (1969) pointed out these analogies in comparing the composition of the lunar surface in Mare Tranquillitatis (as derived from the alpha-scattering experiment on Surveyor V) with those of the ilmenite-rich pyroxene gabbros of the Adirondack Mountains.

Another intriguing possibility is that the titanium-rich basalts on the Moon may have been derived from the partial melting of an amphibole peridotite, a rock type which O'Hara et al. (1970) have suggested as a possible material for the Moon's interior. Amphibole is a rock-forming silicate with a capability for marked titanium concentration; the variety kaersutite can contain up to 10% TiO_2. The Apollo 11 samples can be fairly closely matched with terrestrial kaersutites, as can be seen in the following comparison (A = Apollo sample 10084, analyzed by Wiik and Ojanpera, 1970; B = kaersutite, Iki Islands, Japan, analysed by Aoki, 1959) (* = all Fe as FeO):

	SiO_2	TiO_2	Al_2O_3	FeO	MnO	MgO	CaO	Na_2O	K_2O
A	42.25	7.24	13.83	15.80	0.20	7.97	11.96	0.43	0.13
B	40.36	7.09	13.78	*10.97	0.14	11.08	10.82	2.78	1.34

The similarities are remarkable, except for sodium and potassium, which are very much higher in the terrestrial amphibole than in the lunar material; however, it has already been noted that the lunar rocks are depleted in alkalies in comparison to terrestrial rocks. Derivation of the lunar basalts by the partial melting of an amphibole peridotite would explain the peculiar enrichment of titanium unaccompanied by enrichment in the related elements vanadium, chromium, and manganese, since these elements do not concentrate in amphiboles. The mode of occurrence of kaersutite in terrestrial rocks indicates that it may be an important phase in the upper mantle of the Earth (Mason, 1968); the peculiar composition of the Apollo 11 basalts suggests that a comparable mineral may be a significant phase in the mantle of the Moon.

CHAPTER 8

IMPLICATIONS FOR LUNAR HISTORY

INTRODUCTION

It is clearly premature to propound a comprehensive theory for the origin and evolution of the Moon at this stage in the space program. Nevertheless, the Apollo missions have established some very significant facts that are fundamental for elucidating this basic problem. A long list of such facts could be developed, but the following seem most important. The surface material is of lunar origin, with only a very minor meteoritic increment. The rocks are igneous in origin, although impact has strongly modified some of the igneous rocks and has been the main agent in generating the lunar regolith. The lunar surface lacks water and organic matter. Igneous differentiation (by partial melting, fractional crystallization, gravitational separation, and possibly liquid immiscibility) has taken place. The rocks are all very old; some are older than any terrestrial rocks, and may be as old as any in the solar system.

ORIGIN OF LUNAR SURFACE FEATURES

Lunar exploration by manned and unmanned spacecraft, although limited in the area covered, has already sharpened our understanding of the surface features of the Moon. The maria are not, as Gilvarry (1960) and others have maintained, basins floored by sediments of ancient seas; nor are they, as Gold (1955) has argued, filled with dust eroded from the highlands. They are flooded by basaltic lavas, probably in the form of a large number of thin flows. Although no Apollo landings have yet been made in the highlands, exotic rock fragments in the mare collections support the thesis, already suggested by the Surveyor VII analysis near Tycho, that they are largely made up of plagioclase-rich rocks. Structurally and chemically, the highlands and maria are analogous to the continents and ocean basins on Earth: the highlands are probably made up of lighter feldspathic rocks poor in magnesium and iron, while the maria contain denser rocks rich in these elements. The topographic expression on the Moon is similar to the isostatic adjustment of terrestrial continents riding high in a denser substratum and possibly derives from the same causes.

The origin and history of the maria were, of course, extensively discussed even before the current era of space exploration. The development of a mare may involve two distinct and possibly independent events, the formation of the topographical depression and its partial filling. It has been generally accepted that the maria basins are giant craters excavated in the lunar surface by the impacts of large meteorites at an early period in the Moon's history. The circular outlines of many of the maria have been taken as evidence of an impact origin. However, other origins are possible. On Earth it appears that the ocean basins have been formed by the very slow upwelling and lateral spread of material beneath the mid-oceanic ridges. This upwelling of upper mantle material appears to account for the very active basaltic volcanism along the ocean ridges. It might be argued that the maria were produced at various times in the lunar past by such upwellings. These upwellings, presumably convection cells driven by thermally created density differences, have the effect of sweeping along the continents. They thus provide a ready mechanism for removing continental crust from oceanic areas and for creating the highland-maria topographic contrasts. If such convection cells did play a role in forming the maria, it would appear that they were much more episodic on the Moon than on the Earth and that they occurred mainly, if not entirely, in the Moon's ancient past. In addition, they would also have to differ from their terrestrial counterparts in producing circular, rather than elongate, basins.

The asymmetrical distribution of the maria on the Moon has been ascribed to a concentration of basaltic liquid on the present near side, with an antipodal bulge of light crust on the far side (Anderson et al., 1970). Impacting bodies destroyed the thin near-side crust, exposing large areas of basaltic liquid, whereas similar impacts on the thick far-side crust would form deep craters in solid rock. Whatever their origin, however, the mare basins were filled by a process that took a finite and possibly lengthy period of time. The available data on the time of solidification of mare surface rocks is about 3.7×10^9 years for those from the Apollo 11 site, and 1.7–2.7×10^9 years for those from Apollo 12.

The surface of the maria is covered by a layer of fragmental debris, the lunar regolith. At the Apollo 11 and Apollo 12 sites the regolith is evidently comparatively thin, up to a few meters deep. Both sites are pockmarked with abundant craters of all diameters up to several hundred meters. Many of these craters have sharply raised rims and shallow flat floors, or floors with central humps. The floors of these craters are believed to be excavated in the more cohesive rocky substratum underlying the regolith, specifically the lunar lava flows. Some of these craters have probably been formed by incoming meteorites; some are probably secondary craters formed by the impact of fragments ejected from elsewhere on the Moon.

The low viscosity of the parent magmas of the Apollo 11 basalts has important implications for the origin of certain lunar surface features. Even though the lunar gravity is but one-sixth that of the Earth, the very low viscosity and higher density of the lunar lavas suggest that they would flow twice as far as terrestrial basalts on similar slopes. Such very fluid behavior is in keeping with the kind of volcanism required to build broad, flat floors, rather than volcanic edifices. With such low viscosity, the buildup of areas around vents is unlikely. Rather, if extruded in large quantities in a short time, such lavas would effectively smooth out topography by ponding in any low areas. This low viscosity also favors the formation of lava tubes and thus supports the idea that many of the lunar sinous rilles are in fact collapsed lava tubes (Oberbeck et al., 1969). The formation of lava tubes requires the buildup of an area by repeated flows or ponding followed by the "freezing over" of the main lava channel. The rille then develops after the rate of flow decreases, the lava drains from beneath the roof, and the roof collapses. The parent magmas of the Apollo 12 rocks may have been somewhat more viscous than those of the Apollo 11 samples. Nevertheless, they probably had viscosities comparable, and most likely no higher, than those of terrestrial flood basalts. Thus they too fit the idea of infilling of a mare by repeated outpourings of fluid basic and ultrabasic lavas, and of the creation of rilles by lava tube formation and subsequent collapse.

INTERNAL STRUCTURE OF THE MOON

The moment of inertia of the Moon is very close to that of a homogeneous sphere, which indicates that the Moon's internal structure is quite unlike that of the Earth. In particular, this implies the absence of a heavy metallic core; if such a core is present it must be less than 2% of the total mass. Further evidence for the absence of a core is the negligible magnetic field of the Moon: measurements from orbiting spacecraft have demonstrated that the Moon has less than 0.001 the magnetic field of the Earth.

The bulk density, 3.34 g/cm^3, is a significant constraint on speculations regarding the internal structure of the Moon. The mean density of the

Apollo 11 rocks is about 3.3 g/cm^3, that of the Apollo 12 rocks slightly less, because of their lower ilmenite content. Thus the density of the surface rocks, at least in the maria, is not appreciably less than that of the Moon as a whole. This clearly leaves little possibility for a general increase in density with depth, as occurs in the Earth. It is also clear that the composition of the Moon's interior must be different from that of the surface rocks; Ringwood and Essene (1970) have demonstrated experimentally that, at pressures above about 10 kb, equivalent to a depth of about 200 km on the Moon, the Apollo 11 basalts would transform to a garnet-rich rock with a density of about 3.7 g/cm^3. The interior of the Moon cannot be as alumina-rich as the surface rocks, since under quite moderate pressures alumina combines with other common oxides to give garnet, which has a minimum density of 3.6 g/cm^3. Ringwood and Essene favor pyroxenite as the dominant material in the interior of the Moon, with approximate composition $FeO/(FeO + MgO) = 0.25$, $Al_2O_3 \sim 4\%$, $CaO \sim 3\%$; this composition would give rise to lunar basalts by partial melting, and would provide the observed lunar density and moment of inertia. Other possibilities, suggested by O'Hara et al. (1970) and others, include peridotite (olivine + pyroxene) and amphibole peridotite. O'Hara and his co-workers point out that amphibole peridotite could be equivalent to carbonaceous chondrite composition and also provide the required density for the Moon. The presence of amphibole might imply combined water in the lunar interior, since amphibole is a ferromagnesian silicate with OH groups in the structure, equivalent to 1–2% combined water; however, amphiboles with O, Cl, or F in place of OH are possible. As pointed out in the preceding chapter, the presence of an amphibole such as kaersutite in the lunar mantle would provide a satisfactory explanation for many of the geochemical peculiarities of the lunar basalts.

THE NATURE OF MASCONS

An unexpected dividend of the Orbiter program of unmanned spacecraft was the discovery of mascons—localized mass concentrations below the Moon's surface (Muller and Sjogren, 1968). Mascons are localized below the maria, but not all maria have mascons; they are associated with the circular maria, and not with those having irregular outlines. In the short time since their discovery, mascons have stimulated a remarkable amount of speculation as to their nature and origin. Some of the theories are:

1. Each mascon represents an asteroidal-sized body, probably a large nickel-iron meteorite, which caused its associated mare by impact (Muller and Sjogren, 1968).

2. Mascons reflect high-density lava filling the mare basins, aided by isostatic adjustments that caused the mare bottom to sink (Baldwin, 1968).

3. Mascons are accumulations of heavy minerals, probably titanium-iron oxides, at the base of lava lakes which filled the mare basins (O'Hara et al., 1970).

4. Mascons are bodies of eclogite (garnet + clinopyroxene + rutile), the high-pressure (>12 kb at 1100°C), high-density equivalent of the low-pressure basaltic composition (clinopyroxene + plagioclase + ilmenite) (Ringwood and Essene, 1970).

Thus the origin of mascons is quite controversial, and will only be resolved by additional information. Large meteorites impacting the Earth suffer explosive disruption, and their fragments are scattered over the surrounding landscape, so one would not expect a crater-forming meteorite to survive as a mass at the bottom of a lunar crater. However, it is conceivable that Moon-impacting meteorites were traveling relatively slowly and would plow into its surface with little disintegration. Baldwin's theory has been further developed by Wood et al. (1970) from consideration of the Apollo 11 results. They point out that at hydrostatic equilibrium the mare basin would only be partly filled with solid basalt; further extrusion of lava on the mare surface would constitute a genuine increase of mass to that area of the Moon and give rise to a positive gravity anomaly (Fig. 8-1). It seems unlikely that the mare basins were ever lava lakes deep

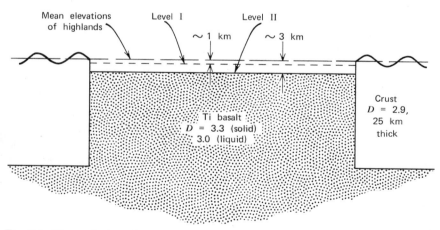

Fig. 8-1. The evolution of a mascon, as suggested by Wood et al. (1970). A mare basin, excavated in a plagioclase-rich crust of density 2.9, is filled by liquid basalt ($D = 3.0$) to level I, determined by buoyant equilibrium, but the surface falls to level II after it crystallizes and becomes denser ($D = 3.3$). Repeated eruptions of lava (in hydrostatic equilibrium with subsurface magma sources) on to this surface will produce a positive gravity anomaly, that is, a mascon.

enough to produce the accumulations of heavy minerals postulated by O'Hara and co-workers. The explanation of Ringwood and Essene would place the mascons at great depths (of the order of 200 km), whereas they appear to be near-surface phenomena (Cook, 1970).

THE ORIGIN OF THE MOON

One of the principal objectives of lunar exploration is, of course, to determine how and when the Moon came into being as a satellite of the Earth. This problem has exercised men's minds for many generations. Currently there appear to be four major hypotheses, each with some support in the scientific community:

1. The fission hypothesis, the Moon having separated from the Earth's mantle.
2. The double-planet hypothesis, whereby the Earth and the Moon formed in close proximity by accretion from similar parental material.
3. The hypothesis that the Moon formed elsewhere in the solar system and was then captured by the Earth.
4. The hypothesis that the Moon formed by the coalescence of a ring of planetesimals that once surrounded the Earth.

Serious objections have been raised to all these hypotheses, and no consensus has yet developed. Nevertheless, the great increase in our knowledge of the Moon has considerably restricted the boundary conditions, and a notable refinement in our understanding of lunar genesis is now possible.

The fission hypothesis is associated with the name of Sir George Darwin, who first propounded it in 1880. He suggested that, at an early stage in Earth history, tidal forces raised a huge bulge that became dynamically unstable and was then ejected from the Earth to form the Moon. This hypothesis provided an elegant explanation for the Moon's density, which is similar to the uncompressed density of the Earth's mantle. However, the hypothesis fell into disfavor after Jeffreys (1930) raised serious objections to the dynamics of the process. It has since been revived by several investigators, for example, Ringwood (1960), O'Keefe (1966, 1969), and Wise (1963, 1969), who link the fission of the Moon from the Earth with the rapid separation of the core from the mantle. Additional interest in the fission hypothesis has arisen from the possible lunar origin of tektites; tektite composition is so similar to terrestrial surface material that a lunar origin for these objects would support the idea of the derivation of the Moon from the outer layers of the Earth. However, a lunar origin of tektites seems unlikely, and the dynamical difficulties of the fission hypothesis still remain. The results of the Apollo missions are the strongest argument against the fission hypothesis. Lunar and terrestrial basalts show compositional differences of such magnitude as to preclude their derivation from a common source material, that is, the mantle of the Earth. Ringwood and Essene (1970) point out that the Apollo 11 source region has been depleted in sodium, potassium, rubidium, lead, and zinc, compared to the terrestrial basalt source region, by factors of 3 to 10, and is also markedly depleted in siderophile elements such as nickel, copper, and gallium.

The double-planet hypothesis has been developed by Kuiper (1954, 1959), who suggests that the Earth-Moon system is analogous to a binary star system, the two bodies having formed in close proximity by accretion from similar parent material. However, the low density of the Moon relative to the Earth immediately raises a serious problem. This could be overcome by postulating a physical fractionation process during accretion that concentrated metallic iron in the Earth, but a plausible mechanism is not obvious. Alternatively, it has been argued that the Earth and the Moon have similar composition, but the material of the Moon is more oxidized and hydrated, thereby lowering its density to the required value. Here again the evidence from the Apollo missions is in direct contradiction, the lunar material being anhydrous.

The capture hypothesis was propounded by Urey, who has discussed it in a number of papers (1960, 1962, 1965). He considered the Moon to be a rare survivor from a generation of primary bodies in the solar system, from which the present terrestrial planets were formed by complex mechanisms of aggregation and fractionation. Much of the argument for this hypothesis revolved around the Fe/Si abundance ratio, which was thought to be much lower in the Sun than in the Earth and the chondritic meteorites. Urey thus explained the low density of the Moon by a low Fe/Si ratio, similar to that of the Sun and presumably to that of the primitive solar nebula. However, recent investigations have shown that the solar iron abundance is much greater than previously estimated, so that this argument is no longer valid. The capture theory also suffers from its low intrinsic probability; capture processes are highly unlikely events.

The hypothesis of the origin of the Moon by the coalescence of a ring of planetesimals originated with Öpik (1955). He visualized this ring as being somewhat analogous to the rings of Saturn but much more massive; it circled the Earth at a distance of 5–8 Earth radii. He based his hypothesis on the study of lunar craters and tidal evolution but did not explain the origin of the planetesimal ring or the cause of the Moon's low density. Ringwood (1966) developed Opik's hypothesis, with special attention to the geochemical and geophysical implications. He proposes that, in the later stages of accretion of the Earth, a massive primitive atmosphere was present and the temperature was high enough selectively to vaporize a substantial proportion of the accreting silicates. This primitive atmosphere was subsequently dissipated as the Sun passed through a highly active phase. With the disruption and escape of the primitive atmosphere, the accompanying expansion and cooling caused the precipitation of the silicates as a ring of planetesimals circling the Earth. This ring was inherently unstable and coalesced to form the Moon.

Of the above hypotheses, that of Öpik-Ringwood provides the most plausible explanations for many lunar riddles, especially those revealed by compositional data on the lunar rocks. The low density of the Moon and its essential homogeneity as revealed by its moment of inertia are compatible with its being an aggregate of silicate planetesimals. The extremely low abundance of the siderophile elements, with the apparent absence of a metallic core, indicates the removal of some iron and practically all the more noble metals prior to accretion—these elements were presumably left behind in the Earth. Volatile elements, such as bismuth, mercury, zinc, cadmium, thallium, lead, sulfur, selenium, chlorine, and bromine, are significantly depleted with respect to their probable abundance in the solar system; their loss is readily understandable if the planetesimal ring was subjected to high temperatures. On the other hand, highly refractory elements (scandium, titanium, strontium, yttrium, zirconium, hafnium, niobium, tantalum, barium, and the rare earths) are markedly enriched in the lunar rocks, again indicating a fractionation of the elements by a high-temperature phase in the development of the planetesimals.

The Öpik-Ringwood hypothesis is actually a close relative of the classic fission theory, since it maintains that the material now in the Moon was originally part of the Earth. The basic difference in the more recent theory is that, instead of having been part of the Earth's mantle, the moon is believed to have been derived from material contained in a massive primitive atmosphere.

REFERENCES

Agrell, S. O., J. H. Scoon, I. D. Muir, J. V. P. Long, J. D. C. McConnell, and A. Peckett (1970). "Mineralogy and petrology of some lunar samples." *Science,* 167, 583–586.

Albee, A. L., D. S. Burnett, A. A. Chodes, O. J. Eugster, J. C. Huneke, D. A. Papanastassiou, F. A. Podosek, G. P. Russ, H. G. Sanz, F. Tera, and G. J. Wasserburg (1970). "Ages, irradiation history, and chemical composition of lunar rocks from the Sea of Tranquillity." *Science,* 167, 463–466.

Anderson, A. T., A. V. Crewe, J. R. Goldsmith, P. B. Moore, R. C. Newton, E. J. Olsen, J. V. Smith, and P. J. Wyllie (1970). "Petrologic history of the Moon suggested by petrography, mineralogy, and crystallography." *Science,* 167, 587–589.

Anderson, D. L., and R. A. Phinney (1967). "Early thermal history of the terrestrial planets," in *Mantles of the Earth and Terrestrial Planets.* Edited by S. K. Runcorn. Wiley, New York, pp. 114–138.

Annell, C., and A. Helz (1970). "Emission spectrographic determination of trace elements in lunar samples." *Science,* 167, 521–523.

Aoki, K. (1959). "Petrology of alkali rocks of the Iki Islands and Higashimatsuura district, Japan." *Sci. Rept. Tohoku Univ.,* Ser. 3, 6, 261–310.

Arrhenius, G., S. Asunmaa, J. I. Drever, J. Everson, R. W. Fitzgerald, J. Z. Frazer, H. Fujita, J. S. Hanor, D. Lal, S. S. Liang, D. Macdougall, A. M. Reid, J. Sinkankas, and L. Wilkening (1970). "Phase chemistry, structure and radiation effects in lunar samples." *Science,* 167, 659–661.

Bailey, J. C., P. E. Champness, A. C. Dunham, J. Esson, W. S. Fyfe, W. S. MacKenzie, E. F. Stumpfl, and J. Zussman (1970). "Mineralogical and petrological investigations of lunar samples." *Science,* 167, 592–594.

Baldwin, R. B. (1963). *The Measure of the Moon.* University of Chicago Press, 488 pp.

Baldwin, R. B. (1968). "Lunar mascons: another interpretation." *Science,* 162, 1407–1408.

Barnes, V. E. (1961). "Tektites." *Sci. Amer.,* 205, No. 5, 58–65.

Brown, G. M., C. H. Emeleus, J. G. Holland, and R. Phillips (1970). "Petrographic, mineralogic and X-ray fluorescence analysis of lunar igneous-type rocks and spherules." *Science,* 167, 599–601.

Cameron, E. N. (1970). "Opaque minerals in lunar samples." *Science,* 167, 623–625.

Cassidy, W. A., B. Glass, and B. C. Heezen (1969). "Physical and chemical properties of Australasian microtektites." *J. Geophys. Res.,* 74, 1008–1025.

Chao, E. C. T., O. B. James, J. A. Minkin, J. A. Boreman, E. D. Jackson, and C. B. Raleigh (1970). "Petrology of unshocked crystalline rocks and shock effects in lunar rocks and minerals." *Science,* 167, 644–647.

Chayes, F. (1956). *Petrographic modal analysis.* Wiley, New York, 113 pp.

Compston, W., P. A. Arriens, M. J. Vernon, and B. W. Chappell (1970). "Rubidium-strontium chronology and chemistry of lunar material." *Science,* 167, 474–476.

Cook, A. H. (1970). "The Earth as a planet." *Nature,* 226, 18–20.

Corlett, M., and P. H. Ribbe (1967). "Electron probe microanalysis of minor elements in plagioclase feldspars." *Schweizerische Mineralogische und Petrographische Mitteilungen,* 47, 317–332.

Derham, C. J., and J. E. Geake (1964). "Luminescence of meteorites." *Nature,* 201, 62–63.

Doell, R. R., C. S. Grommé, A. N. Thorpe, and F. E. Senftle (1970). "Magnetic studies of lunar samples." *Science,* 167, 695–697.

Douglas, J. A. V., M. R. Dence, A. G. Plant, and R. J. Traill (1970). "Mineralogy and deformation in some lunar samples." *Science,* 167, 594–597.

Duke, M. B., and L. T. Silver (1967). "Petrology of eucrites, howardites and mesosiderites." *Geochim. Cosmochim. Acta,* 31, 1637–1666.

Duke, M. B., C. C. Woo, M. L. Bird, G. A. Sellers, and R. B. Finkelman (1970). "Lunar soil: size distribution and mineralogical constituents." *Science,* 167, 648–650.

Ehmann, W. D., and J. W. Morgan (1970). "Oxygen, silicon, and aluminum in lunar samples by 14 MeV neutron activation." *Science,* 167, 528–530.

Engel, A. E. J., and C. G. Engel (1970). "Lunar (Apollo 11) rock compositions and some interpretations." *Science,* 167, 527–528.

Epstein, S., and H. P. Taylor (1970). "$^{18}O/^{16}O$, $^{30}Si/^{28}Si$, D/H, and $^{13}C/^{12}C$ studies of lunar rocks and minerals." *Science,* 167, 533–536.

Evans, B. W., and J. G. Moore (1968). "Mineralogy as a function of depth in the prehistoric Makaopuhi tholeiitic lava lake, Hawaii," *Contr. Mineral Petrol.,* 17, 85–115.

Evans, H. T. (1970). "Lunar troilite crystallography." *Science,* 167, 621–623.

Fleischer, R. L., E. L. Haines, R. E. Hanneman, H. R. Hart, J. S. Kasper, E. Lifshin, R. T. Woods, and P. B. Price (1970). "Particle track, X-ray, thermal, and mass spectrometric studies of lunar material." *Science,* 167, 568–571.

Fluscher, R. L., P. B. Price, and R. M. Walker (1969). "Nuclear tracks in solids." *Sci. Amer.,* 220, No. 6, 30–39.

Fredriksson, K., J. Nelen, W. G. Melson, E. P. Henderson, and C. A. Anderson (1970). "Lunar glasses and micro-breccias: properties and origin." *Science,* 167, 664–666.

French, B. M., and N. M. Short, Eds. (1968). "Shock metamorphism of natural materials." *Mono Book Corp.,* Baltimore, 644 pp.

Friedman, I., J. R. O'Neil, L. H. Adami, J. D. Gleason, and K. Hardcastle (1970). "Water, hydrogen, deuterium, carbon, carbon-13, and oxygen-18 content of selected lunar material." *Science,* 167, 538–541.

Frondel, C., C. Klein, J. Ito, J. C. Drake (1970). "Mineralogy and composition of lunar fines and selected rocks." *Science,* 167, 681–683.

Funkhouser, J. G., O. A. Schaeffer, D. D. Bogard, and J. Zähringer (1970). "Gas analysis of the lunar surface." *Science,* 167, 561–563.

Ganapathy, R., R. R. Keays, J. C. Laul, and E. Anders (1970). "Trace elements in

Apollo 11 lunar rocks: implications for meteorite influx and origin of Moon." *Geochim. Cosmochim. Acta, Suppl.* 1, 1117–1142.

Gast, P. W., and N. J. Hubbard (1970). "Abundance of alkali metals, alkaline and rare earths, and strontium-87/strontium-86 ratios in lunar samples." *Science,* 167, 485–487.

Gay, P., G. M. Bancroft, and M. G. Bown (1970). "Diffraction and Mössbauer studies of minerals from lunar soils and rocks." *Science,* 167, 626–628.

Geake, J. E. (1964). "Lunar luminescence." *Nature,* 204, 866–867.

Gilvarry, J. J. (1960). "Origin and nature of lunar surface features." *Nature* 188, 886–891.

Gold, T. (1955). "The lunar surface." *Mon. Not. R. Astron. Soc.,* 115, 585–595.

Gold, T. (1969). "Apollo 11 observations of a remarkable glazing phenomenon on the lunar surface." *Science,* 165, 1345–1349.

Goldstein, J. I., and R. E. Ogilvie (1965). "A re-evaluation of the iron-rich portion of the Fe-Ni system." *Trans. Metall. Soc. Am. Inst. Mining Eng.,* 233, 2083–2087.

Gordon, G. G., M. Osawa, K. Randle, R. Beyer, D. Y. Jerome, D. J. Lindstrom, M. R. Martin, S. M. McKay and T. L. Steinborn (1970). "Instrumental neutron activation analyses of lunar specimens." *Science,* 167, 497–499.

Green, T. H. (1969). "High-pressure experimental studies on the origin of anorthosite." *Canadian J. Earth Sci.,* 6, No. 3, 427–440.

Haggerty, S. E., F. R. Boyd, P. M. Bell, L. W. Finger and W. B. Bryan (1970). "Iron-titanium oxides and olivine from 10020 and 10071." *Geochim. Cosmochim. Acta,* in press.

Hallimond, A. F. (1919). "The crystallography of vogtite: a metasilicate of iron, calcium, manganese, and magnesium, from acid steel-furnace slags." *Mineral. Mag.,* 18, 368–372.

Hargraves, R. B., and W. E. Perkins (1969). "Investigations of the effect of shock on natural remanent magnetism." *J. Geophys. Res.,* 74, 2576–2589.

Haskin, L. A., R. O. Allen, P. A. Helmke, T. P. Paster, M. R. Anderson, R. L. Korotev, and K. A. Zweifel (1970). "Rare earths and other trace elements in Apollo 11 lunar samples." *Geochim. Cosmochim. Acta, Suppl.* 1, 1213–1231.

Helsley, C. E. (1970). "Magnetic properties of lunar dust and rock samples." *Science,* 167, 692–695.

Herr, W., U. Herpers, B. Hess, B. Skerra, and R. Woelfle (1970). "Determination of manganese-53 by neutron activation and other miscellaneous studies on lunar dust." *Science,* 167, 747–749.

Hess, H. H., and E. P. Henderson (1949). "The Moore County meteorite: a further study with comment on its primordial environment." *Am. Mineral.,* 34, 494–507.

Hintenberger, H., H. W. Weber, H. Voshage, H. Wänke, F. Begemann, E. Vilscek, and F. Wlotzka (1970). "Rare gases, hydrogen, and nitrogen: concentrations and isotopic composition in lunar material." *Science,* 167, 543–545.

Jacobs, J. A. (1963). *The earth's core and geomagnetism.* Macmillan, New York.

Jeffreys, H. (1930). "The resonant theory of the origin of the Moon." *Mon. Not. R. Astron. Soc.,* 91, 169–173.

Johannsen, A. (1937). *A descriptive petrography of the igneous rocks.* Vol. 3. University of Chicago Press.

Kaplan, I. R., and J. W. Smith (1970). "Concentration and isotopic composition of carbon and sulfur in Apollo 11 lunar samples." *Science,* 167, 541–543.

Kaula, W. A. (1969). "The gravitational field of the Moon." *Science,* 166, 1581–1588.

Keays, R. R., R. Ganapathy, J. C. Laul, E. Anders, G. F. Herzog, and P. M. Jeffery (1970). "Trace elements and radioactivity in lunar rocks: implications for meteorite infall, solar-wind flux, and formation conditions of Moon." *Science,* 167, 490–493.

Keil, K., M. Prinz, and T. E. Bunch (1970). "Mineral chemistry of lunar samples." *Science,* 167, 597–599.

King, E. A., M. F. Carman, and J. C. Butler (1970). "Mineralogy and petrology of coarse particulate material from lunar surface at Tranquillity Base." *Science,* 167, 650–652.

Kopal, Z., and T. W. Rackham (1964). "Excitation of lunar luminescence by solar flares." *Nature,* 201, 239–241.

Kuiper, G. P. (1954). "On the origin of the lunar surface features." *Proc. Nat. Acad. Sci.,* 40, 1096–1112.

Kuiper, G. P. (1959). "The Moon." *J. Geophys. Res.,* 64, 1713–1719.

Kushiro, I., Y. Nakamura, H. Haramura, and S. Akimoto (1970). "Crystallization of some lunar mafic magmas and generation of rhyolitic liquid." *Science,* 167, 610–612.

Larochelle, A., and E. J. Schwarz (1970). "Magnetic properties of lunar sample 10048-22." *Science,* 167, 700–701.

Lindsley, D. H., and C. W. Burnham (1970). "Pyroxferroite: stability and X-ray crystallography of synthetic $Ca_{0.15}Fe_{0.85}SiO_3$." *Science,* 168, 364–368.

Lindsley, D. H., and J. L. Munoz (1969). "Subsolidus relations along the join hedenbergite-ferrosilite." *Am. J. Sci.,* 267-a, 295–324.

Lovering, J. F., and D. Butterfield (1970). "Neutron activation analysis of rhenium and osmium in Apollo 11 lunar material." *Geochim. Cosmochim. Acta, Suppl.* 1, 1351–1355.

LSPET (Lunar Sample Preliminary Examination Team) (1969). "Preliminary examination of lunar samples from Apollo 11." *Science,* 165, 1211–1227.

LSPET (Lunar Sample Preliminary Examination Team) (1970). "Preliminary examination of lunar samples from Apollo 12." *Science,* 167, 1325–1339.

Malkus, W. V. R. (1968). "Precession of the Earth as the cause of geomagnetism." *Science,* 160, 259–264.

Marti, K., G. W. Lugmair, and H. C. Urey (1970). "Solar wind gases, cosmic ray spallation products, and the irradiation history." *Science,* 167, 548–550.

Mason, B. (1962). *Meteorites.* Wiley, New York, 274 pp.

Mason, B. (1968). "Kaersutite from San Carlos, Arizona, with comments on the paragenesis of this mineral." *Mineral. Mag.,* 36, 997–1002.

Mason, B., K. Fredriksson, E. P. Henderson, E. Jarosewich, W. G. Melson, K. M. Towe, and J. S. White (1970). "Mineralogy and petrography of lunar samples." *Science,* 167, 656–659.

Mason, B., and A. L. Graham (1970). "Minor and trace elements in meteoritic minerals." *Smithsonian Contrib. Earth Sci.,* 3, in press.

Mason, B., and W. G. Melson (1970). "Comparison of lunar rocks with basalts and stony meteorites." *Geochim. Cosmochim. Acta,* in press.

Mason, B., and H. B. Wiik (1966). "The composition of the Barratta, Carraweena, Kapoeta, Moresport, and Ngasir meteorites." *Amer. Mus. Novitates,* 2273, 1–20.

McKay, D. S., W. R. Greenwood, and D. A. Morrison (1970). "Morphology and related chemistry of small lunar particles from Tranquillity Base." *Science,* 167, 654–656.

Melson, W. G., and G. Switzer (1966). "Plagioclase-spinel-graphite xenoliths in metallic iron-bearing basalts, Disko Island, Greenland." *Am. Mineral.,* 51, 664–676.

Melson, W. G., G. T. Thompson, and T. H. Van Andel (1968a). "Volcanism and metamorphism in the mid-Atlantic Ridge, 22° N. latitude." *J. Geophys. Res.,* 73, 5925–5941.

Melson, W. G., E. Jarosewich, E. P. Henderson (1968b). "Metallic phases in terrestrial basalts: implications on equilibria between basic magmas and iron-carbon melts." *Trans. Am. Geophys. Union,* 49, 352 (Abstract).

Melson, W. G., and E. Jarosewich (1967). "St. Peter and St. Paul Rocks: a high-temperature, mantle-derived intrusion." *Science,* 155, 1532–1535.

Miyashiro, A., F. Shido, and M. Ewing (1970). "Crystallization and differentiation in abyssal tholeiites and gabbros from mid-oceanic ridges." *Earth Planetary Sci. Letters,* 7, 361–365.

Moore, C. B., C. F. Lewis, E. K. Gibson, and W. Nichiporuk (1970). "Total carbon and nitrogen abundances in lunar samples." *Science,* 167, 459–497.

Morrison, G. H., J. T. Gerard, A. T. Kashuba, E. V. Gangadharam, A. M. Rothenberg, N. M. Potter, and G. B. Miller (1970). "Multielement analysis of lunar soil and rocks." *Science,* 167, 505–507.

Muller, P. M., and W. L. Sjogren (1968). "Lunar gravity: preliminary estimates from lunar Orbiter." *Science,* 159, 625–627.

Murase, T., and A. McBirney (1970). "Viscosity of lunar lavas." *Science,* 167, 1491–1492.

NASA (1969). "Apollo 11 preliminary science report." NASA SP-214, National Aeronautics and Space Administration, Washington, D. C., 201 pp.

Nagata, T. (1961). *Rock magnetism.* 2nd edition. Maruzen Co., 350 pp.

Nagata, T., Y. Ishikawa, H. Kinoshita, M. Kono, Y. Syono, and R. M. Fisher (1970). "Magnetic properties of the lunar crystalline rock and fines." *Science,* 167, 703–704.

Oberbeck, V. R., W. L. Quaide, and R. Greeley (1969). "On the origin of lunar sinuous rilles." *Modern Geol.,* 1, 75–80.

O'Hara, M. J., G. M. Biggar, and S. W. Richardson (1970). "Experimental petrology of lunar material: the nature of mascons, seas, and the lunar interior." *Science,* 167, 605–607.

O'Keefe, J. A. (1966). "The origin of the Moon and the core of the Earth," in *The Earth-Moon system,* B. G. Marsden and A. G. W. Cameron, Eds. Plenum Press, New York, pp. 224–233.

O'Keefe, J. A. (1969). "Origin of the Moon." *J. Geophys. Res.,* 74, 2758–2767.

O'Keefe, J. A. (1970). Tektite glass in Apollo 12 sample. *Science,* 168, 1209–1210.

O'Keefe, J. A., P. D. Lowman, and W. S. Cameron (1967). "Lunar ring dikes from Lunar Orbiter I." *Science,* 155, 77–79.

Olsen, E. (1969). "Pyroxene gabbro (anorthosite association): similarity to Surveyor V lunar analysis." *Science,* 166, 401–402.

Onuma, N., R. N. Clayton, T. K. Mayeda (1970). "Oxygen isotope fractionation between minerals and an estimate of the temperature of formation." *Science,* 167, 536–538.

Öpik, E. J. (1955). "The origin of the Moon." *Irish Astron. J.,* 3, 245–248.

Peck, L. C., and V. C. Smith (1970). "Quantitative chemical analysis of lunar samples." *Science,* 167, 532.

Philpotts, J. A., and C. C. Schnetzler (1970). "Potassium, rubidium, strontium, barium, and rare-earth concentrations in lunar rocks and separated phases." *Science,* 167, 493–495.

Quaide, W., T. Bunch, and R. Wrigley (1970). "Impact metamorphism of lunar surface materials." *Science,* 167, 671–673.

Quaide, W. L., and V. R. Oberbeck (1969). "Geology of the Apollo landing sites." *Earth Sci. Rev.,* 5, 255–278.

Ramdohr, P., and A. El Goresy (1970). "Opaque minerals of the lunar rocks and dust from Mare Tranquillitatis." *Science,* 167, 615–618.

Reed, G. W., S. Jovanovic, and L. H. Fuchs (1970). "Trace elements and accessory minerals in lunar samples." *Science,* 167, 501–503.

Ringwood, A. E. (1960). "Some aspects of the thermal evolution of the Earth." *Geochim. Cosmochim. Acta,* 20, 241–259.

Ringwood, A. E. (1966). "Chemical evolution of the terrestrial planets." *Geochim. Cosmochim. Acta,* 30, 41–104.

Ringwood, A. E., and E. Essene (1970). "Petrogenesis of lunar basalts and the internal constitution and origin of the Moon." *Science,* 167, 607–610.

Roedder, E., and P. W. Weiblen (1970a). "Silicate liquid immiscibility in lunar magmas, evidenced by melt inclusions in lunar rocks." *Science,* 167, 641–644.

Roedder, E., and P. W. Weiblen (1970b). "Silicate immiscibility found in lunar rocks." *Geotimes,* 15, No. 3, 10–13.

Runcorn, S. K. (1967). "Convection in the planets," in *Mantles of the Earth and terrestrial planets,* S. K. Runcorn, Ed. Wiley, New York, pp. 513–524.

Runcorn, S. K., D. W. Collinson, W. O'Reilly, A. Stephenson, N. N. Greenwood, and M. H. Battey (1970). "Magnetic properties of lunar samples." *Science,* 167, 697–699.

Schminke, H. V. (1967). "Stratigraphy and petrography of 4 upper Yakima basalt flows in south central Oregon." *Bull. Geol. Soc. Am.,* 78, 1385–1422.

Shoemaker, E. M., and R. J. Hackman (1962). "Stratigraphic base for a lunar time scale," in *The Moon,* Z. Kopal and Z. K. Mikhailov, Eds. Academic Press, London and New York, pp. 289–300.

Shoemaker, E. M., M. H. Hait, G. A. Swann, D. L. Schleicher, D. H. Dahlem, G. G. Schaber, and R. L. Sutton (1970). "Lunar regolith at Tranquillity Base." *Science,* 167, 452–454.

Short, N. M. (1970). "Evidence and implications of shock metamorphism in lunar samples." *Science,* 167, 673–677.

Silver, L. T. (1970). "Uranium-thorium-lead isotope relations in lunar materials." *Science,* 167, 468–471.

Simpson, P. R., and S. H. U. Bowie (1970). "Quantitative optical and electron-probe studies of the opaque phases." *Science,* 167, 619–621.

Skinner, B. J. (1970). "High crystallization temperatures indicated for igneous rocks from Tranquillity Base." *Science,* 167, 652–654.

Smales, A. A., D. Mapper, and K. F. Fouché (1967). "The distribution of some trace elements in iron meteorites as determined by neutron activation." *Geochim. Cosmochim. Acta,* 31, 673–720.

Smales, A. A., D. Mapper, M. S. W. Webb, R. K. Webster, and J. D. Wilson (1970). "Elemental composition of lunar surface material." *Science,* 167, 509–512.

Spencer, L. J. (1933). "Meteorite iron and silica-glass from the meteorite-craters of

REFERENCES

Henbury (central Australia) and Wabar (Arabia)." *Mineral. Mag.*, 23, 387–404.

Stacey, F. D., J. F. Lovering, and L. G. Parry (1961). "On the magnetic properties of meteorites." *J. Geophys. Res.*, 66, 1523.

Strangway, D. W., E. E. Larson, and G. W. Pearce (1970). "Magnetic properties of lunar samples." *Science*, 167, 691–693.

Tatsumoto, M., and J. N. Rosholt (1970). "Age of the Moon: an isotopic study of uranium-thorium-lead systematics of lunar samples." *Science*, 167, 461–463.

Turkevich, A. L., E. J. Franzgrote, and J. H. Patterson (1967). "Chemical analysis of the Moon at the Surveyor V landing site." *Science*, 158, 635–637.

Turkevich, A. L., E. J. Franzgrote, and J. H. Patterson (1969). "Chemical composition of the lunar surface in Mare Tranquillitatis." *Science*, 165, 277–279.

Urey, H. C. (1959). "Primary and secondary objects." *J. Geophys. Research*, 64, 1721–1737.

Urey, H. C. (1960). "Lines of evidence in regard to the composition of the Moon," in *Space Research*, H. K. Kallman-Bÿl, Ed. North Holland Publishing Co., Amsterdam, pp. 1114–1120.

Urey, H. C. (1962). "Origin and history of the Moon," in *Physics and Astronomy of the Moon*, Z. Kopal, Ed. Academic Press, New York, Chapter 13.

Urey, H. C. (1965). "Meteorites and the Moon." *Science*, 147, 1262–1265.

Urey, H. C. (1967). "Parent bodies of the meteorites and the origin of chondrules." *Icarus*, 7, 350–359.

Vinogradov, A. P., Y. A. Surkov, G. M. Chernov, F. F. Kirnozov, and G. B. Nazarkina (1966). "Measurement of gamma-radiation from the surface of the Moon by the cosmic station Luna 10." *Geochim. Internat.*, 3, 707–715.

Von Engelhardt, W., J. Arndt, W. F. Muller, and D. Stoffler (1970). "Shock metamorphism in lunar samples." *Science*, 167, 669–670.

Walter, L. S., and R. N. Clayton (1967). "Oxygen isotopes: experimental vapor fractionation and variations in tektites." *Science*, 156, 1357–1358.

Wänke, H. (1966). Der Mond als Mutterkörper der Bronzit-Chondrite." *Z. Naturforsch.*, 21a, 93–110.

Wänke, H., F. Begemann, E. Vilcsek, R. Rieder, F. Teschke, W. Born, M. Quijano-Rico, H. Voshage, and F. Wlotzka (1970). "Major and trace elements and cosmic-ray produced radioisotopes in lunar samples." *Science*, 167, 523–526.

Wasson, J. T., and P. A. Baedēcker (1970). "Ga, Ge, In, Ir and Au in lunar, terrestrial and meteoritic basalts." *Geochim. Cosmochim. Acta, Suppl.* 1, 1741–1750.

Weill, D. F., I. S. McCallum, Y. Bottinga, M. J. Drake, and G. A. McKay (1970). "Petrology of a fine-grained igneous rock from the Sea of Tranquillity." *Science*, 167, 635–638.

Wiik, H. B., and P. Ojanpera (1970). "Chemical analyses of lunar samples 10017, 10072, and 10084." *Science*, 167, 531–532.

Wise, D. U. (1963). "On origin of the Moon by fission during formation of the Earth's core." *J. Geophys. Res.*, 68, 1547–1554.

Wise, D. U. (1969). "Origin of the Moon from the Earth: some new mechanisms and comparisons." *J. Geophys. Res.*, 74, 6034–6045.

Wood, J. A., J. S. Dickey, U. B. Marvin, and B. N. Powell (1970). "Lunar anorthosites and a geophysical model of the Moon." *Geochim. Cosmochim. Acta., Suppl.* 1, 965–988.

Yagi, K., and K. Onuma (1967). "The join $CaMgSi_2O_6$-$CaTiAl_2O_6$ and its bearing on the titanaugites." *J. Fac. Sci. Hokkaido Univ.*, Ser. IV, 13, 463–483.

INDEX

Achondrites, enstatite, 113
  and lunar rocks, 108–113
Adirondack Mountains, 153
Age, of lunar rocks, 76–77, 98–99
Agrell, S. O., 32, 58, 68, 70
Albee, A. L., 50, 77
Albite, 38
Aldrin, E. E., 15, 16, 46, 82
Alkali basalts, 104
Alpha particle analysis, 10
Aluminum, in carbonaceous chondrites, 117
  in eucrites, 117
  in lunar rocks, 117, 130
  in pyroxenes, 36
Amphibole, in Apollo 11 rocks, 32
  in lunar interior, 153–154, 158
Amphibole peridotite, 153, 158
Anderson, A. T., 48, 64, 68, 101, 104, 156
Anderson, D. L., 77, 149
Andesine, 95
Andesitic glass, 68
Annell, C., 125, 139, 140
Anorthite, 38, 89
Anorthosites, 95–96, 103, 153
Antiferromagnetism, 70
Antimony, in carbonaceous chondrites, 118
  in eucrites, 118
  in lunar rocks, 118, 142
Aoki, K., 153
Apatite, 33, 50, 55, 131, 148
Apennine Mountains, 2, 4, 9, 27, 28
Apollo 11 rocks, elemental abundances in, 117–119
Apollo 12 fines, elemental abundances in, 117–119
Apollo Lunar Surface Experiments Package (ALSEP), 20
Apollo program, 14–31
Aragonite, 32
Aristarchus, 26, 30
Argon, in lunar rocks, 123–124
Armalcolite, 33, 46–47, 56, 64, 66
Armstrong, N. A., 15, 46, 82, 83
Arrhenius, G., 32, 41, 139
Arsenic, in carbonaceous chondrites, 118
  in eucrites, 118
  in lunar rocks, 118, 137
Atmophile elements, 121
Augite, 34, 35, 36, 65, 105, 106

Baddeleyite, 33, 49, 139
Baedecker, P. A., 137, 146
Bailly, 4
Baldwin, R. B., 8, 77, 159
Barium, in carbonaceous chondrites, 118
  in eucrites, 118
  in lunar rocks, 118, 120, 143
  in potash feldspar, 40
Barnes, V. E., 114
Basalts, as lunar rocks, 18, 19, 21, 25, 100–108
  of Columbia River Plateau, 8, 65, 102
  of Disko Island, 104
  Hawaiian, 106
  ocean-ridge, 101
  source regions for, 107
Basanites, 106
Base surge deposits, 86
Beryllium, in carbonaceous chondrites, 117
  in eucrites, 117

171

in lunar rocks, 117, 125
Bismuth, in carbonaceous chondrites, 119
  in lunar rocks, 119, 147
Boron, in carbonaceous chondrites, 117
  in eucrites, 117
  in lunar rocks, 117, 125
Bowie, S. H. U., 46
Bromine, in carbonaceous chondrites, 118
  in eucrites, 118
  in lunar rocks, 118, 138, 153
Bronzite, 106
Brown, G. M., 103, 106
Burnham, C. W., 104
Butterfield, D., 146
Bytownite, 38

Cadmium, in carbonaceous chondrites, 118
  in eucrites, 118
  in lunar rocks, 118, 141
Calcium, in carbonaceous chondrites, 117
  in eucrites, 117
  in lunar rocks, 117, 132
Cameron, E. N., 103
Carbon, in carbonaceous chondrites, 117
  in eucrites, 117
  in lunar rocks, 117, 125–126
  isotopic composition, 126
Carbonaceous chondrites, elemental abundances in, 117–119, 151
Cassidy, W. A., 115
Censorinus, 26, 27
Cerium, in carbonaceous chondrites, 119
  in eucrites, 119
  in lunar rocks, 119, 120, 143
Cesium, in carbonaceous chondrites, 118
  in eucrites, 118
  in lunar rocks, 118, 142–143
  in potash feldspar, 40
Chalcophile elements, 121, 149, 152
Chalcopyrite, 32
Chao, E. C. T., 53, 81, 90, 96
Chlorapatite, 50
Chlorine, in apatite, 50
  in carbonaceous chondrites, 117
  in eucrites, 117
  in lunar rocks, 117, 131, 153
Chondrites, bronzite, 108
  carbonaceous, 97
  lunar origin of, 108
Chondrules, 109

Chromite, 33, 48
Chromium, in carbonaceous chondrites, 117
  in eucrites, 117
  in lunar rocks, 117, 134
  in pyroxenes, 36
Classification, of elements, 121
  of lunar rocks, 60–62
Clavius, 4
Clayton, R. N., 90
Clinohypersthene, 34
Clinopyroxene, 57, 64, 66, 89
Cobalt, in carbonaceous chondrites, 117
  in eucrites, 117
  in lunar rocks, 117, 135
  in metal particles, 43
Cobra Head, 30
Cohenite, 33, 46, 126
Collins, M., 46, 92
Columbia River Plateau, 8, 102
Command module, 14
Compston, W., 139, 140
Contact metamorphism, 57
Cook, A. H., 160
Copernican Period, 5
Copernicus, 2, 5, 9, 26, 27, 28
Copper, as lunar mineral, 33, 46
  in carbonaceous chondrites, 118
  in eucrites, 118
  in lunar rocks, 118, 136
Corlett, M., 38
Cosmic dust, 123
Cosmic rays, 84, 123
Cosmogenic gases, 123–124
Crater chains, 1
Craters, lunar, 1, 4, 80, 157
Cristobalite, 33, 41–43, 69, 110, 111
Crystallization, of lunar magmas, 63–67
Crystal settling, 65
Cumulates, 63, 96
Curie temperature, 70
Curium, 10

Darwin, G., 160
Density, lunar magmas, 62
Derham, C. J., 113
Deuterium, in lunar rocks, 122
Differentiation, of lunar magmas, 63
Disko Island basalts, 104
Doell, R. R., 73
Dolerites, 20, 25, 61, 105

Douglas, J. A. V., 87
Domes, lunar, 27
Duke, M. B., 83, 90, 108
Dysanalyte, 49, 55
Dysprosium, in carbonaceous chondrites, 119
　in eucrites, 119
　in lunar rocks, 119

Ehmann, W. D., 127, 130
Elemental abundances, in carbonaceous chondrites, 117–119
　in eucrites, 117–119
　in lunar rocks, 117–119
　quantization of, 120
Elements, atmophile, 121
　chalcophile, 121, 149, 152
　lithophile, 121
　rare-earth, 143–145
　siderophile, 121, 149, 152, 162
El Goresy, A., 32, 43, 46, 49
Enstatite, 35, 89
Enstatite achondrites, 113
Engel, A. E. J., 109
Engel, C. G., 109
Epstein, S., 65, 90, 128
Eratosthenes, 2, 9
Eratosthenian Period, 5
Erbium, in carbonaceous chondrites, 119
　in eucrites, 119
　in lunar rocks, 119
Essene, E., 64, 66, 78, 107, 158, 159, 160, 161
Eucrites, and lunar rocks, 12, 108–112
　elemental abundances in, 117–119
Europium, in carbonaceous chondrites, 119
　in eucrites, 119
　in lunar rocks, 119, 144–145
Evans, B. W., 106
Evans, H. T., 43
Extravehicular activity (EVA), 15, 16, 18, 20, 26

Faults, on Moon, 4
Fayalite, 37, 41
Feldspar, plagioclase, 37–39
　potash, 39–40, 68
Ferroaugite, 34, 35, 36
Ferrobasalts, 64, 101
Ferromagnetism, 70
Ferrosilite, 70, 71, 89
Fines, lunar, 20, 59, 83

Fission tracks, 85
Fission hypothesis, 160–162
Flamsteed, 6
Fleischer, R. L., 85
Flood basalts, 102
Fluorapatite, 128
Fluorine, in apatite, 128
　in carbonaceous chondrites, 117
　in eucrites, 117
　in lunar rocks, 117, 128, 153
Fra Mauro region, 25, 26
Fredriksson, K., 36, 88, 95
French, B. M., 86
Friedman, I., 122
Frondel, C., 46, 89
Funkhouser, J. G., 84, 123

Gabbro-anorthosite complexes, 103
Gabbros, 61, 101
Gadolinium, in carbonaceous chondrites, 119
　in eucrites, 119
　in lunar rocks, 119
Galileo, 1, 5
Gallium, in carbonaceous chondrites, 118
　in eucrites, 118
　in lunar rocks, 118, 136–137
Ganapathy, R., 142, 147
Garnet, 66, 158
Gast, P. W., 142
Gay, P., 32
Geake, J. E., 113
Germanium, in carbonaceous chondrites, 118
　in eucrites, 118
　in lunar rocks, 118, 137
Gilvarry, J. J., 155
Glass, B., 91
Glasses, lunar, 20, 22, 67–69, 82, 84, 87–93
Gold, in carbonaceous chondrites, 119
　in eucrites, 119
　in lunar rocks, 119, 146
Gold, T., 155
Goldschmidt, V. M., 121
Goldstein, J. I., 46
Goles, G. G., 129, 133
Graham, A. L., 125
Grain size, igneous rocks, 54, 61
Granulite texture, 57, 58
Graphite, 32

Hackman, R. J., 5
Hadley's Rille, 4, 26, 28

Hafnium, in baddeleyite, 49
  in carbonaceous chondrites, 119
  in eucrites, 119
  in lunar rocks, 119, 145
  in perovskite, 49
Haggerty, S. E., 44, 45, 47, 48
Hallimond, A. F., 37
Hargraves, R. B., 76
Haskin, L. A., 78, 138, 144
Hedenbergite, 35, 89
Helium, in lunar rocks, 123–124
Helsley, C. E., 72, 73, 74, 76
Helz, A., 125, 139, 140
Hematite, in lunar rocks, 32
Henderson, E. P., 109
Herodotus, 30
Herr, W., 146
Hess, H. H., 109
Highlands, lunar, 1, 25, 156
Hintenberger, H., 123, 127
Holmium, in carbonaceous chondrites, 119
  in eucrites, 119
  in lunar rocks, 119
Hornblende, 104
Hornfels, 96
Howardites, and lunar rocks, 108–112
Hubbard, N. J., 142
Hydrogen, in lunar rocks, 122–123
Hyginus Rille, 9, 26, 31
Hypersthene, 34, 106

Igneous rock, 52
Ilmenite, 19, 21, 25, 33, 34, 38, 41, 47, 52, 65, 66, 89, 103, 139, 153
Imbrian Period, 5
Immiscibility, liquid, 62
Impact metamorphism, 86–87
Impacts, secondary, 95
Increment, meteoritic, 96–98, 120, 135, 149
Indium, in carbonaceous chondrites, 118
  in eucrites, 118
  in lunar rocks, 118, 141
Inert gases, in lunar rocks, 123–124
Iodine, in carbonaceous chondrites, 118
  in eucrites, 118
  in lunar rocks, 118, 142
Iridium, in carbonaceous chondrites, 119
  in eucrites, 119
  in lunar rocks, 119, 146
Iron, in carbonaceous chondrites, 117
  in eucrites, 117
  in lunar rocks, 32, 43–46, 55, 70, 104, 117, 134
Iron-sulfur melt, 55, 104
Isostasy, on Moon, 28
Isotopes, carbon, 126
  oxygen, 64, 65, 127–128
  silicon, 128
  sulfur, 131

Jeffreys, H., 160
Johannsen, A., 60, 101

Kaersutite, 153–154, 158
Kamacite, 45
Kaplan, I. R., 126, 131
Kapoeta howardite, 109–112
Karooite, 64, 66
Kaula, W. M., 9
Keays, R. R., 97, 98, 136, 137, 138, 140, 141, 142, 146, 147, 149
Keil, K., 47, 49, 50, 68
Kepler crater, 7, 9, 113
King, E. A., 86
Kopal, Z., 113
Krypton, in lunar rocks, 123–124
Kuiper, G. P., 161
Kushiro, I., 68

Labradorite, 38
Lanthanum, in carbonaceous chondrites, 118
  in eucrites, 118
  in lunar rocks, 118
Larochelle, A., 73
Laser Ranger Retroreflector (LRRR), 17
Lava tubes, on Moon, 157
Lead, in carbonaceous chondrites, 119
  in eucrites, 119
  in lunar rocks, 119, 147
  isotopic composition, 153
Lindsley, D. H., 104
Liquid immiscibility, 67–69
Liquidus, lunar magmas, 64, 106
Lithium, in carbonaceous chondrites, 117
  in eucrites, 117
  in lunar rocks, 117, 123
Lithophile elements, 121
Littrow area, 25
Lovering, J. F., 146
Luminescence, lunar, 113
Luna program, 8, 10
Lunar craters, 1, 4, 80

INDEX   175

Lunar Excursion Module (LM), 14
Lunar fines, composition, 117–119
  grain size, 83
Lunar glasses, 87–93
Lunar gravity, 8
Lunar highlands, 1, 25, 156
Lunar magmas, crystallization sequence, 63–67
  density, 62–63
  differentiation, 63
  origin, 77–79
  temperatures, 62, 106
  viscosity, 62–63, 157
Lunar magnetic field, 72
Lunar mantle, 107
Lunar mapping, 5
Lunar Receiving Laboratory (LRL), 18, 20
Lunar regolith, 17, 23, 80–85
Lunar rocks, age, 8, 76–77, 98–99
  analyses, 59, 96, 101, 115, 153
  chemical composition, 57–60, 96, 101, 115, 117–119
  classification, 60–62
  magnetic properties, 69–76
  and meteorites, 108–115
  mineralogical composition, 52–55
  texture, 54
Lunar Sample Preliminary Examination Team (LSPET), 18, 23, 32, 34, 36, 46, 53, 57, 58, 60, 77, 78, 84, 115, 123, 126, 133, 139, 143, 148
Lutetium, in carbonaceous chondrites, 119
  in eucrites, 119
  in lunar rocks, 119

McBirney, A., 62
McKay, D. S., 86, 88, 90
Magmas, lunar, 52
Magnesium, in carbonaceous chondrites, 117
  in eucrites, 117
  in lunar rocks, 117, 129
Magnetic field, of Moon, 76, 157
Magnetism, rock, 69–76
Magnetite, 32
Malkus, W. V. R., 72
Manganese, in carbonaceous chondrites, 117
  in eucrites, 117
  in lunar rocks, 117, 134
  in pyroxenes, 36
Manned Spacecraft Center (MSC), 18

Mantle, lunar, 107
Mare Aestuum, 13
Mare Cognitum, 9
Mare Crisium, 9, 13
Mare Fecunditatis, 9
Mare Humboltianum, 13
Mare Humorum, 9, 13
Mare Imbrium, 2, 7, 9, 13, 25, 28
Mare Nectaris, 9, 13
Mare Nubium, 3, 4, 9
Mare Orientale, 13
Mare Serenitatis, 9, 13, 25, 28
Mare Smythii, 13
Mare Tranquillitatis, 9, 10, 11, 12, 15, 27, 62, 80, 109, 133, 153
Mare Vaporum, 31
Maria, distribution, 9, 13
  origin, 155–157
Marius Hills, 9, 26, 27, 29
Marti, K., 123
Marvin, U. B., 49
Mascons, 13, 28, 158–160
Maskelynite, 39
Mason, B., 35, 44, 48, 55, 78, 95, 102, 109, 110, 125, 154
Melabasalts, 60, 101
Meladolerites, 101
Melting temperatures, of impact glasses, 88
  of crystalline igneous rocks, 64
Melson, W. G., 55, 78, 101, 102, 104
Mercury, in carbonaceous chondrites, 119
  in lunar rocks, 119, 146–147
Mesostasis, 55
Mesosiderites, 108
Metamorphism, contact, 57
  shock, 39
Meteorites, achondritic, 78
  and lunar rocks, 108–115
  carbonaceous chondrites, 97, 117–119, 151
  eucritic, 12, 108–112, 117–119
Meteoritic increment, 96–98, 120, 135, 149
Mica, 32
Microbreccias, 20, 22, 59, 85–86, 111–112
Microcraters, 93–95
Microgabbros, 61
Microtektites, 113
Mineralogy, normative, 61
Miyashiro, A., 101, 103
Molybdenum, in carbonaceous chondrites, 118
  in lunar rocks, 118, 140

Moon, age, 8
  core, 76
  density, 8, 157
  diameter, 8
  gravity, 8
  heat sources, 77
  internal structure, 157–158
  magnetic field, 76, 157
  mass, 8
  mantle, 107
  moment of inertia, 157
  origin, 160–162
  volume, 8
Moore, C. B., 125, 126, 127
Moore, J. G., 106
Moore County eucrite, 109–112
Morgan, J. W., 127, 130
Morrison, G. H., 125, 128, 131, 137, 138, 140, 141, 142, 145
Mössbauer spectra, 41
Mösting, C., 26, 27
Mountains, lunar, 1
Muller, P. M., 158
Munoz, J. L., 104
Murase, T., 62

Nagata, T., 69, 70, 71, 72, 73
Neodymium, in carbonaceous chondrites, 119
  in eucrites, 119
  in lunar rocks, 119
Neon, in lunar rocks, 123–124
Nickel, in carbonaceous chondrites, 118
  in eucrites, 118
  in lunar rocks, 118, 135
  in metal particles, 43
Nickel-iron, 33, 43–46, 70, 89, 91, 94, 97, 149
Niobium, in carbonaceous chondrites, 118
  in lunar rocks, 118, 140
Nitrides, in lunar material, 127
Nitrogen, in carbonaceous chondrites, 117
  in eucrites, 117
  in lunar rocks, 117, 127
Noble gases, in lunar rocks, 123–124
Nuclear tracks, 85

Oberbeck, V. R., 86, 157
Oceanus Procellarum, 6, 9, 10, 20, 27, 29, 30, 93
Ogilvie, R. E., 46

O'Hara, M. J., 64, 153, 158, 159, 160
Ojanpera, P., 153
O'Keefe, J. A., 6, 115, 160
Olivine, 25, 33, 41, 52, 57, 64, 65, 66
Olsen, E., 153
Onuma, K., 36
Onuma, N., 65, 113, 127
Ophitic texture, 65, 104, 105
Öpik, E. J., 161, 162
Orbiter program, 11–13
Organic compounds, in lunar materials, 125, 126
Orthopyroxene, 36
Osmium, in carbonaceous chondrites, 119
  in eucrites, 119
  in lunar rocks, 119, 146
Oxygen, in carbonaceous chondrites, 117
  in eucrites, 117
  in lunar rocks, 117, 127–128
  fagacity, 104
  isotopic composition, 64, 90, 127
  partial pressure, 152

Palladium, in carbonaceous chondrites, 118
  in lunar rocks, 118, 140
Paramagnetism, 70
Partial fusion, 77
Passive Seismic Experiment Package (PSEP), 16
Peck, L. C., 11, 59
Peridotites, 61, 158
Period, Copernican, 5
  Eratosthenian, 5
  Imbrian, 5
  Pre-Imbrian, 5
Perkins, W. E., 76
Perovskite, 33, 49, 55
Philpotts, J. A., 139, 144
Phinney, R. A., 77, 149
Phosphorus, in carbonaceous chondrites, 117
  in eucrites, 117
  in lunar rocks, 117, 130–131
Picotite, 106
Pigeonite, 34, 36, 66, 106, 107
Plagioclase, 19, 21, 33, 37–39, 52, 64, 89, 105, 110
Platinum, in carbonaceous chondrites, 119
  in lunar rocks, 119, 146
Potash feldspar, 33, 39–40, 68, 132, 138
Potassium, in carbonaceous chondrites, 117

INDEX 177

in eucrites, 117
in lunar rocks, 117, 119, 131–132
Praseodymium, in carbonaceous chondrites, 119
in eucrites, 119
in lunar rocks, 119
Pre-Imbrian Period, 5
Procellarum, Oceanus, 6, 9, 10, 20, 27, 29, 30, 93
Pseudobrookite, 47
Pyroxene, 19, 21, 33–36, 38, 52, 56, 106, 110
Pyroxferroite, 33, 34, 35, 36–37, 55, 69, 104
Pyroxmangite, 36, 37

Quadie, W. L., 46, 86
Quartz, 33, 41–43, 61

Rackham, T. W., 113
Radiation effects, in lunar rocks, 84–85
Radiogenic gases, in lunar rocks, 123–124
Ramdohr, P., 32, 43, 46, 49
Ranger program, 9
Rare-earth elements, in apatite, 50
in carbonaceous chondrites, 118–119
in eucrites, 118–119
in lunar rocks, 118–119, 143–145
in whitlockite, 50
Rare gases, in lunar rocks, 123–124
Rays, lunar, 1
Reed, G. W., 128, 131, 138, 142, 146
Regolith, lunar, 17, 23, 80–85
Remanent magnetism, 70, 72–74
Rhenium, in carbonaceous chondrites, 119
in eucrites, 119
in lunar rocks, 119, 146
Rhodium, in carbonaceous chondrites, 118
in lunar rocks, 118, 140
Ribbe, P. H., 38
Ridges, wrinkle, 1
Rilles, lunar, 1, 29, 157
Rima Hadley, 4, 28, 31
Ringwood, A. E., 64, 66, 78, 107, 158, 159, 160, 161
Roedder, E., 64, 67, 68
Rosholt, J. N., 77, 147, 148
Rubidium, in carbonaceous chondrites, 118
in eucrites, 118
in lunar rocks, 118, 120, 138
in potash feldspar, 40

Runcorn, S. K., 73, 75
Ruthenium, in carbonaceous chondrites, 118
in lunar rocks, 118, 140
Rutile, 33, 47, 49, 66

Samarium, in carbonaceous chondrites, 119
in eucrites, 119
in lunar rocks, 119
Sanidine, 40, 61
Scandium, in carbonaceous chondrites, 117
in eucrites, 117
in lunar rocks, 117, 133
Schnetzler, C. C., 139, 144
Schreibersite, 33, 46, 131
Schroeter's Valley, 26, 30, 31
Schwarz, E. J., 73
Scoon, J. H., 59
Selenium, in carbonaceous chondrites, 118
in eucrites, 118
in lunar rocks, 118, 138
Service module, 14, 25
Shergotty meteorite, 39
Shock damage, 86
Shock deformation, 80
Shock lithification, 86
Shock waves, magnetic effects, 76
Shoemaker, E. M., 5, 83
Short, N. M., 86
Siderophile elements, 121, 149, 152, 162
Silica minerals, 41–43
Silicon, in carbonaceous chondrites, 117
in eucrites, 117
in lunar rocks, 117, 130
isotopic composition, 90, 127
Silver, in carbonaceous chondrites, 118
in lunar rocks, 118, 141
Silver, L. T., 77, 108, 147
Simpson, P. R., 46
Sinus Medii, 10, 11
Sjogren, W. L., 158
Skaergaard intrusion, 103
Skinner, B. J., 104
Smales, A. A., 136, 137
Smith, J. W., 126, 131
Smith, V. C., 11
Sodium, in carbonaceous chondrites, 117
in eucrites, 117
in lunar rocks, 117, 129
in pyroxenes, 36
Soil, lunar, 17, 81
Solar flares, 84

INDEX

Solar wind, 17, 84, 123, 127, 149
Solidus, lunar magmas, 64, 106
Spallogenic gases, 123–124
Spencer, L. J., 89
Spherules, glass, 88
  nickel-iron, 88, 94
Spinel, 33, 47, 49, 106
St. Paul's Rocks, 104
Straight Wall, 3, 4, 9
Strangway, D. W., 70, 72, 73
Strontium, in carbonaceous chondrites, 118
  in eucrites, 118
  in lunar rocks, 118, 139
Suess, F. E., 113
Suevites, 101
Sulfur, in carbonaceous chondrites, 117
  in eucrites, 117
  in lunar rocks, 117, 131
  isotopic composition, 131
Surveyor program, 9, 10
Switzer, G., 104

Taenite, 45
Tantalum, in carbonaceous chondrites, 119
  in eucrites, 119
  in lunar rocks, 119, 145
Tatsumoto, M., 77, 147, 148
Taylor, H. P., 65, 90, 128
Tektites, 108, 113–115, 160
Tellurium, in carbonaceous chondrites, 118
  in eucrites, 118
  in lunar rocks, 118, 142
Temperature, Curie, 70
Terbium, in carbonaceous chondrites, 119
  in eucrites, 119
  in lunar rocks, 119
Terrae, 1
Thallium, in carbonaceous chondrites, 119
  in eucrites, 119
  in lunar rocks, 119, 147
Thermal history, Moon, 77
Thorium, in carbonaceous chondrites, 119
  in eucrites, 119
  in lunar rocks, 119, 120, 148
Thulium, in carbonaceous chondrites, 119
  in eucrites, 119
  in lunar rocks, 119
Tin, in carbonaceous chondrites, 118
  in lunar rocks, 118, 141–142
Titanium, in carbonaceous chondrites, 117
  in eucrites, 117
  in lunar rocks, 59, 117, 133
  in pyroxenes, 36
Towe, K. M., 93
Tranquillity Base, 77, 81, 83, 84, 93
Tridymite, 33, 41–43, 110, 111
Troctolites, 61
Troilite, 33, 43, 55, 131
Tungsten, in carbonaceous chondrites, 119
  in lunar rocks, 119, 145
Turkevich, A. L., 10, 12, 109, 133, 151
Tycho, 9, 10, 11, 26, 27, 155

Ulvöspinel, 33, 47–48
Uranium, in carbonaceous chondrites, 119
  in eucrites, 119
  in lunar rocks, 119, 120, 148–149
Urey, H. C., 108, 161

Vanadium, in carbonaceous chondrites, 117
  in eucrites, 117
  in lunar rocks, 117, 133–134
Vesicles, 52, 92
Vinogradov, A. P., 10
Viscosity, lunar magmas, 62
Vogtite, 37
Volatilization, of impact glasses, 88, 90
Volcanism, in lunar history, 51
Von Engelhardt, W., 88, 90
Vugs, 52

Walter, L. S., 46, 90
Wänke, H., 108, 125, 131, 137, 145
Wasson, J. T., 137, 146
Water, in lunar rocks, 58, 65, 122–123
Weiblen, P. W., 64, 67, 68
Weill, D. F., 62
Whitlockite, 33, 50, 55, 131, 148
Wiik, H. B., 109, 153
Wise, D. U., 160
Wood, J. A., 159
Wrinkle ridges, 1, 25, 29

Xenon, in lunar rocks, 123–124

Yagi, K., 36
Ytterbium, in carbonaceous chondrites, 119
  in eucrites, 119
  in lunar rocks, 119
Yttrium, in carbonaceous chondrites, 118
  in eucrites, 118
  in lunar rocks, 118, 139

Zinc, in carbonaceous chondrites, 118
   in eucrites, 118
   in lunar rocks, 118, 136
Zircon, 33, 49–50, 139
Zirconium, in carbonaceous chondrites, 118
   in eucrites, 118
   in ilmenite, 41, 139
   in lunar rocks, 118, 120, 139
   in perovskite, 49